AF676364

Blockchain Technology for Enhancing Supply Chain Performance and Reducing the Threats Arising from the COVID-19 Pandemic

Blockchain Technology for Enhancing Supply Chain Performance and Reducing the Threats Arising from the COVID-19 Pandemic

Editors

Kamalakanta Muduli
Rakesh Raut
Balkrishna Eknath Narkhede
Himanshu Shee

MDPI • Basel • Beijing • Wuhan • Barcelona • Belgrade • Manchester • Tokyo • Cluj • Tianjin

Editors
Kamalakanta Muduli
Papua New Guinea
University of Technology
Papua New Guinea

Rakesh Raut
National Institute of
Industrial Engineering
(NITIE)
India

Balkrishna Eknath Narkhede
National Institute of
Industrial Engineering
(NITIE)
India

Himanshu Shee
Victoria University
Australia

Editorial Office
MDPI
St. Alban-Anlage 66
4052 Basel, Switzerland

This is a reprint of articles from the Special Issue published online in the open access journal *Sustainability* (ISSN 2071-1050) (available at: https://www.mdpi.com/journal/sustainability/special_issues/Blockchain_Supply_Chain_COVID-19).

For citation purposes, cite each article independently as indicated on the article page online and as indicated below:

LastName, A.A.; LastName, B.B.; LastName, C.C. Article Title. *Journal Name* **Year**, *Volume Number*, Page Range.

ISBN 978-3-0365-3731-3 (Hbk)
ISBN 978-3-0365-3732-0 (PDF)

© 2022 by the authors. Articles in this book are Open Access and distributed under the Creative Commons Attribution (CC BY) license, which allows users to download, copy and build upon published articles, as long as the author and publisher are properly credited, which ensures maximum dissemination and a wider impact of our publications.
The book as a whole is distributed by MDPI under the terms and conditions of the Creative Commons license CC BY-NC-ND.

Contents

About the Editors

Kamalakanta Muduli, is presently working as an Associate Professor in the Department of Mechanical Engineering, Papua New Guinea University of Technology, Lae, Morobe Province, Papua New Guinea. He obtained a PhD from the School of Mechanical Sciences, IIT Bhubaneswar, Orissa, India. He obtained a Master's degree in industrial engineering. Dr Muduli has over 15 years of academic experience in universities in India and Papua New Guinea. Dr Muduli is a recipient of the ERASMUS+ KA107 award provided by the European Union. He has published 57 papers in peer-reviewed international journals, most of which are indexed in Clarivate analytics, Scopus and listed in ABDC and more than 25 papers in National and International Conferences. He has been guest-edited Special Issues in journals and books that were approved for publication by Taylor and Francis, MDPI, CRC Press, Wiley scrivener and Apple Academic Press. Dr Muduli has tutored three PhD students. His current research interests include materials science, manufacturing, sustainable supply chain management, and Industry 4.0 applications in operations and supply chain management. Dr Muduli is a fellow of the Institution of Engineers India. He is also a senior member of the Indian Institution of Industrial Engineering and member of ASME.

Rakesh D. Raut is an Associate Professor of Operations and Supply Chain Management at the National Institute of Industrial Engineering, Mumbai. Rakesh D. Raut received his Post-Doctoral Fellow (LGPP) from EPFL, Switzerland, and a Fellowship (Ph.D., Operations and Supply Chain Management) from the National Institute of Industrial Engineering (NITIE), Mumbai. He holds an M. Tech (Mechanical) and BE (Production) Degree from Nagpur University. He has more than nine years of work experience in industry and academic institutions. Before joining NITIE, he worked as an Assistant Professor at IMT, Nagpur, Dubai, and Symbiosis Centre for Management and Human Resource Development (SCMHRD), Pune. His research interests include Industry 4.0, cloud computing adoption, cloud-IoT adoption, Big Data analytics, greensustainable supply chain management, He has co-authored several research items in peer-reviewed international journals, such as *International Journal of Production Research, International Journal of Production Economics, Transportation Research Part E: Logistics and Transportation Review, Journal of Cleaner Production, Renewable Energy, Annals of Operations Research, Business Strategy and the environment*, etc.

Balkrishna Eknath Narkhede is a professor of Operations and Supply Chain Management at the National Institute of Industrial Engineering Mumbai, INDIA. Before this, he served at VJTI Mumbai, AIT Bangkok, DTE Mumbai. He has been awarded a D.Sc. (science and technology) specializing in mechanical engineering, as well as a Ph.D. (mechanical engineering), M.Tech. (metallurgical engineering), B.E. (production engineering) and FDPM (general management) from the Indian Institute of Management Indore. He has successfully guided 12 PhD Scholars. He is widely published in International Journals of repute, including *IEEE Transactions of Engineering Management, International Journal of Production Research, Renewable & Sustainable Energy Reviews, Journal of Cleaner Production, Journal of Enterprise Information Management, Industrial Management & Data Systems, Annals of OR, Business Strategy & Environment, International Journal of Advanced Manufacturing Technology, International Journal of Manpower, International Journal of Productivity & Performance Management, Cleaner Technologies and Environmental Policy, OPSEARCH Journal, Renewable Energy, Benchmarking: An International Journal, Journal of Manufacturing Technology & Management, Sustainable Production & Consumption, Journal of Environmental Management, Journal of System & Information Technology*, etc.

He has 2726 citations, an h-index of 30, RG Score of 38.81, 3 patents, and 2 books to his name. His research interests include operations and supply chain management, the circular economy and sustainability.

Himanshu Shee is currently an associate -rofessor at Victoria University, specialising in operations, supply chains and logistics management. With a good track record in research, publications and higher-degree research supervision, he has been published in journals such as the *Int Journal of Logistics Management (IJLM), SCM: an Int. Journal, Int journal of physical distribution and logistics management (IJPDLM), Australian journal of Info systems (AJIS), Int. Journal of Logistics: Research & Applications, IIMB Management Review, Asian Academy of Management Journal, IJGBC, JMO*, and *JTM*. His interdisciplinary research interests include both SCM and emerging communication technology applications in supply chains, including cloud technology, IoT, block chain, BDA, sustainable supply chain management, strategic logistics outsourcing, buyer–supplier dyadic relationships and supply chain innovation. A recipient of the Vice-Chancellor's Award for Excellence in Research and Research Training (supervision), Dr. Shee has significantly contributed to the research, including a reputed international conference presentation at POMS, EUROMA, Academy of Management (AOM), ANZAM-OM and IS-DSI. He received the Best Paper Award in Technology, Innovation & Supply Chain Management stream, as well as the 31st ANZAM Excellence Award. Dr. Shee has acted as a reviewer for *IJPDLM, Internal Journal of Logistics Management (IJLM), Operations Management Research (OMR) Journal*, and *Indian Journal of Commerce*, an associate editor of *International Journal of Global Business* and *Competitiveness (IJGBC)*, and a regular reviewer for many reputed journals and conferences. His teaching experience spans across Australia, China, Singapore, Kuala Lumpur, Fiji and India. He has received the Dean's Outstanding Educator award for his scholarship in teaching. At present, he is the Course Chair for Master's of SCM and Graduate Certificates in transport systems.

Preface to "Blockchain Technology for Enhancing Supply Chain Performance and Reducing the Threats Arising from the COVID-19 Pandemic"

The amount of interest in blockchain technology is driving companies to include it in their technology roadmap agenda. Blockchain has been portrayed as the "next-generation Internet", and "foundational technology", with its innovative potential to transform a wide range of business processes in finance to manufacturing, retailing, healthcare, government, etc. This interest was partly triggered due to the paradigm shift in Industry 4.0 and SCM 4.0, where a plethora of technologies are set to be adopted, as they are deemed to fit into existing practices. For example, IoT, Robotics, AI/ML, 3D printing, and blockchain technologies are among the few have been adopted early, as a pilot, or picked up for selective use. Blockchain can offer many functionalities and capabilities, such as data security, improved visibility in supply chain processes, and an enhanced accuracy and availability of information. However, the lack of understanding of business cases and their ability to integrate into existing processes, along with their cost, benefits, and investment returns, has raised questions regarding how they could be scaled up in a supply chain for a better performance. There is concern regarding how to adopt this technology into current practices.

The supply chain has been identified as one of the non-financial areas (its application is not limited to financial transactions) where the ledger can be used for the storage, authentication and dissemination of transaction records throughout the supply chain. Once a copy of the transaction is recorded, all other copies in a supply chain are updated in real-time, and are ideally immutable. As the supply chain needs close monitoring for optimal functioning at each nodal point, disparate systems can provide limited transparency and information visibility, leading to poor business intelligence under uncertainty. Blockchain technology is perceived to address these pain points in logistics, document frauds, counterfeit medicines, food adulteration and other issues, irrespective of industry. The COVID-19 pandemic context is an example, where goods, information and fund flows were disrupted due to the multiple restrictions. Authors argue that blockchain technology is ideal for reducing the threats of disruption arising from the COVID-19 pandemic.

This edited book aims to collate the thoughts of various authors, who have presented their state-of-the-art views on blockchain technology and its applications in various fields. The contributed papers help to develop an understanding of how blockchain technology can enhance the efficacy of human activities during the pandemic, improve traceability and visibility in automotive supply chains, support food safety and reliability through the digitalisation of food supply chain and increase the performance of next-generation digital supply chains. The authors believe that digital supply chains powered by blockchain technology develop resilience against disruption, and set the groundwork for a more sustainable performance. We hope that you will find this book interesting, useful and exciting, as it broadens the understanding of the emerging blockchain technology and its applications in supply chains across industry sectors. This book could serve as a suitable reference for academics, researchers, and practitioners in industry and consulting.

Kamalakanta Muduli, Rakesh Raut, Balkrishna Eknath Narkhede, and Himanshu Shee
Editors

sustainability

MDPI

Editorial

Blockchain Technology for Enhancing Supply Chain Performance and Reducing the Threats Arising from the COVID-19 Pandemic

Kamalakanta Muduli [1,2,*], Rakesh Raut [3,*], Balkrishna Eknath Narkhede [3,*] and Himanshu Shee [4,*]

1 Mechanical Engineering Department, Papua New Guinea University of Technology, Lae 411, Papua New Guinea
2 Mechanical Engineering Department, CV Raman Global University, Bhubaneswar 752024, India
3 Operations & Supply Chain Management, National Institute of Industrial Engineering (NITIE), Mumbai 400087, India
4 Supply Chain and Logistics Management, College of Business, Victoria University, Melbourne 3000, Australia
* Correspondence: kamalakantam@gmail.com (K.M.); rraut@nitie.ac.in (R.R.); benarkhede1@gmail.com (B.E.N.); himanshu.shee@vu.edu.au (H.S.)

Citation: Muduli, K.; Raut, R.; Narkhede, B.E.; Shee, H. Blockchain Technology for Enhancing Supply Chain Performance and Reducing the Threats Arising from the COVID-19 Pandemic. *Sustainability* **2022**, *14*, 3290. https://doi.org/10.3390/su14063290

Received: 7 March 2022
Accepted: 9 March 2022
Published: 11 March 2022

Publisher's Note: MDPI stays neutral with regard to jurisdictional claims in published maps and institutional affiliations.

Copyright: © 2022 by the authors. Licensee MDPI, Basel, Switzerland. This article is an open access article distributed under the terms and conditions of the Creative Commons Attribution (CC BY) license (https://creativecommons.org/licenses/by/4.0/).

The blockchain is expected to radically alter people's real-time interactions and transactions, culminating in the birth of a new economy in a digital era. Its applications not only go beyond the simple concept of buying and selling in governments and private enterprises, but it also ultimately impacts and revolutionizes how these critical sectors can work [1] in better and more secure way than traditional business. Through the process of tracking and regulatory compliance reporting, blockchain technology improves supply chain performance through traceability, transparency, and trustworthiness.

In a traditional system, logistics services in the corporate environment are divided into the subcategories of product handling, storage, and transportation. This has further been strengthened by the seven delivery rights of logistics: the right product, the right condition, the right time, the right place, the right client, the right cost, and the right number [2]. These seven Rs can only be attained by strategically managing the transactions related to the delivery of commodities. As globalization has encouraged international trade, the material flow from manufacturer to end customer has become more complex. The flow of goods is accompanied by financial and information flows that are not always in sync. The most basic way to share information still requires a lot of paperwork, which can increase the shipping costs from 15% to 20% [3]. As a result, the digitalization of these operations will result in a notable increase in revenue as well as real-time information flow, which will promote customer satisfaction. Many problems, such as cultural and human differences as well as different regulatory rules make it hard to manage risk in global supply chains. Blockchain technology has been perceived to help solve these challenges.

With COVID-19 control protocols in place, human connections are limited, and firms are obliged to interact via virtual platforms. As a result, blockchain technology is becoming more and more popular with businesses because it can cut down on middlemen, make direct supplier–customer transactions easier, eliminate the need for reconciliation, and provide an updated system for tracking assets and making sure that data are accurate [4]. As product compliance and stakeholder needs are being monitored by regulatory authorities, the long-term performance of supply chains depends on quality, pricing, delivery, resilience, and adaptability. Following COVID-19, it is expected that e-commerce will continue to dominate, while last-mile deliveries and pay-per-service will gain traction. During the COVID-19 period, online buying patterns have increased by more than 50% in nations such as India, Vietnam, and China [5]. In this case, blockchain technology, which protects the supply chain process from tampering, cybercrime, and fraud, is likely to become more popular.

During the pandemic, various preventive measures such as social isolation, lockdowns, travel restrictions, and shutdowns have caused interruptions on supply chain activities across businesses, particularly in the health care sector. For example, healthcare providers had to close their outpatient services in order to prevent their employees from being inadvertently exposed to the virus. As a result, online communication channels serve as a bridge between patients and healthcare providers. Many pharmaceutical companies, such as Medplus and Apollo pharma, among others, have attempted to increase their market share through mobile app-based services. However, the widespread adoption of these technologies is mostly dependent on openness and data security. Thus, the use of blockchain technology will help telemedicine to grow in popularity.

Panic buying and irrational food stockpiling occurred due to perceived shortages during lockdowns and shutdowns, labour shortages, and restricted transportation operations. As a result, this interruption to the food supply created an urgent need to establish an innovative and robust food supply chain to prevent food shortages and waste. A significant number of stakeholders are involved in the supply of raw materials, seeds, equipment, finance, harvesting, temporary storage, warehousing, and shipping. Furthermore, regulatory and compliance reporting are critical in controlling the supply chain of perishable vegetables, pharmaceuticals, fruits, consumer goods, and so on. This is where blockchain technology might be able to help people work together more smoothly and make the supply chain more stable.

Blockchain technology can be examined in terms of its application in logistic management, information transmission, and data management in humanitarian relief, where it provides a traceable and responsible way to manage all financial expenses. It reduces lack of faith in companies and fund deviation, hence eliminating the most fundamental reason for the calamitous failure for the majority of supply chains [6]. Furthermore, the various contributors and donors can keep track of their gifts in a credible and transparent manner. Furthermore, by minimizing communication and transitional complications between suppliers and procurement officials, there can be a significant reduction in supply delays, the timely refilling of inventories, and accelerated purchasing of items [7].

Five pieces have been published in this Special Issue.

Sudhanshu Joshi, Manu Sharma, Rashmi Prava Das, Kamalakanta Muduli, Rakesh Raut, Balkrishna Narkhede, Himanshu Shee, and Abhishek Misra wrote the first paper, "Assessing Effectiveness of Humanitarian Activities Against COVID-19 Disruption: The Role of Blockchain-Enabled Digital Humanitarian Network (BT-DHN)". This study aims to identify and provide insights into important aspects that may improve the efficacy of human activities (HAs) during a pandemic. To investigate crucial factors, a systematic literature study was conducted and evaluated by experts using the fuzzy Delphi approach. The fuzzy decision-making trial and laboratory (DEMATEL) approach was used to further evaluate these factors to determine the cause-and-effect link. According to the findings, the most important component during a pandemic is the establishment of a blockchain-enabled digital humanitarian network (BT-DHN). The utilization of digital platforms for real-time information sharing improves the effectiveness of HAs. This study allows stakeholders, politicians, and decision-makers to evaluate these elements during strategic planning for pandemic disruptions.

Nesrin Ada, Manavalan Ethirajan, Anil Kumar, Vimal K.E.K., Simon Peter Nadeem, Yigit Kazancoglu, and Jayakrishna Kanadasamy wrote the second paper, "Blockchain Technology for Enhancing Traceability and Efficiency in the Automobile Supply Chain—A Case Study". This study investigates the issues that the automotive sector has with its supply chain operations. Traceability concerns and waiting times at various supply chain nodes are regarded as key issues affecting overall supply chain efficiency in the automotive supply chain. This paper suggests a novel blockchain-based architecture to improve traceability and to reduce waiting times for the automotive supply chain after examining existing blockchain architectures and their deployment techniques. A hyper ledger fabric-based blockchain architecture was created to manage ownership transfers in incoming and outbound logistics. The simulation results of the proposed hyper ledger fabric-based blockchain architecture show improved item traceability at different supply chain nodes, improving the Inventory Quality Ratio (IQR), and the mean waiting time was reduced at the factory, wholesaler, and retailer, improving overall supply chain efficiency. A blockchain-enabled supply chain is better suited to removing the risks and uncertainties that are associated with the automotive supply chain. The advantages of implementing blockchain technology in the automobile supply chain are also discussed. The established blockchain-based architecture can provide greater visibility into the movement of items and inventory status in automotive supply chains.

Sumit Kumar Rana, Hee-Cheol Kim, Subhendu Kumar Pani, Sanjeev Kumar Rana, Moon-II Joo, Arun Kumar Rana, and Satyabrata Aich wrote the third article, "Blockchain-Based Model to Improve the Performance of the Next-Generation Digital Supply Chain". Emerging technologies affect all parts of the industrial realm in the fourth industrial revolution era. A digital supply chain is used by businesses to track the delivery of their products or supplies. The digital supply chain is still plagued by difficulties such as a lack of provenance, a lack of transparency, and a lack of confidence. One of the developing technologies, blockchain technology, can be linked to the supply chain to address existing difficulties and to increase performance. In this study, a concept for integrating blockchain technology into the supply chain to increase performance is provided. The suggested model maintains the supply chain's traceability, transparency, and trustworthiness by combining the Ethereum blockchain and the interplanetary file system.

Mohammed Alkahtani, Qazi Salman Khalid, Muhammad Jalees, Muhammad Omair, Ghulam Hussain, and Catalin lulian Pruncu wrote the fourth article, "E-Agricultural Supply Chain Management Coupled with Blockchain Effect and Cooperative Strategies". The agriculture industry is largely underdeveloped and requires technological change to ensure food safety and reliability. Blockchain technology is being used in the digital realm to successfully establish sustainable e-agricultural supply chain management (e-Agri-SCM). With the current blockchain breakthroughs in digital marketing, product website design (web design) is critical to streamlining client requirements and supply chain partner expectations. The blockchain effect has been introduced into this research study using web design features in the agricultural supply chain management (Agri-SCM) study. Furthermore, partners in the digital marketing supply chain (DM-SCM) are struggling to discover significant site design element-based blockchain technology to maximize profit. In this work, a cooperative (Co-op) sustainable e-agricultural SCM model is established by taking the web design index and variable demand into account to determine shipments, selling prices, cycle time, and advertisement costs for agricultural products. The application of the fuzzy system addresses the uncertainties in the model caused by intangible web design aspects and fundamental prices, while carbon emissions are also taken into account to provide a greener output. The suggested model is used in real-time by running five different examples based on the mutual shares, demand curves, and advertisement budgets among the participants. Sensitivity analysis is also used to find critical elements influencing total profit. This study's findings include major web design elements (WDEs), such as web visuals, search engine optimization, cyber-security, rapid loading, and navigation, as essentials for digital marketing to persuade clients to buy a product in a global SCM. The numerical results and management insights are useful for managers in terms of profit maximization through cooperative and digital marketing methods to achieve e-Agri-SCM.

Muhammad Nabipour and M. Ali Ulku wrote the fifth paper, "On Deploying Blockchain Technologies in Supply Chain Strategies and the COVID-19 Pandemic: A Systematic Literature Review and Research Outlook". The advent of a new pandemic, known as the COVID-19 pandemic, has impacted numerous supply chain segments (SC). Numerous studies on the subject have been undertaken since the beginning of the pandemic, but the need for a comprehensive review study that identifies the gaps and limitations of existing research as well as opportunities and agendas for future investigations, is obvious. This study aims to add to the content of previous studies by conducting a systematic literature review on blockchain technology (BCT) implementation in supply chain management (SCM) in the context of the COVID-19 pandemic. Using a number of resources, relevant documents were discovered (Scopus, Google Scholar, Web of Science, and ProQuest). Seventy-two publications were selected systematically using the PRISMA approach and were thoroughly assessed based on BCT, methodology, industrial sectors, geographical context, and sustainability context. According to their findings, there is a substantial absence of empirical and quantitative approaches in the literature. The majority of studies did not take into account specific sectors. Furthermore, there are few papers concentrating on the sustainability context, particularly on social and environmental challenges. Furthermore, the majority of the publications assessed did not take into account the geographical setting. The findings show that the deployment of BCT in various sectors is not consistent, and that their utilization is dependent on their services during the COVID-19 pandemic. Furthermore, the focus of research on the effects of the BCT on SCM varies depending on the circumstances of various nations in terms of the implication of the COVID-19 pandemic. The data also reveal that in the context of the COVID-19 pandemic, there is a direct relationship between the deployment of BCT and sustainability considerations such as economic and waste issues. Finally, this study offers research ideas and agendas to help academics and other stakeholders better understand the current literature, find topics that need more research, and start new studies.

Funding: This research received no external funding.

Institutional Review Board Statement: Not applicable.

Informed Consent Statement: Not applicable.

Data Availability Statement: Not applicable.

Conflicts of Interest: The authors declare no conflict of interest.

References

1. Seebacher, S.; Schüritz, R. Blockchain technology as an enabler of service systems: A structured literature review. In *International Conference on Exploring Services Science*; Springer: Cham, Switzerland, 2017; pp. 12–23.
2. Queiroz, M.M.; Fosso Wamba, S. Blockchain adoption challenges in supply chain: An empirical investigation of the main drivers in India and the USA. *Int. J. Inf. Manag.* **2019**, *46*, 70–82. [CrossRef]
3. Lal, R.; Johnson, S. American Family Insurance and the Artificial Intelligence Opportunity. Harvard Business School Case 519-028, October 2018. Available online: https://www.hbs.edu/faculty/Pages/item.aspx?num=55102 (accessed on 4 April 2021).
4. Hughes, L.; Dwivedi, Y.K.; Misra, S.K.; Rana, N.P.; Raghavan, V.; Akella, V. Blockchain research, practice and policy: Applications, benefits, limitations, emerging research themes and research agenda. *Int. J. Inf. Manag.* **2019**, *49*, 114–129. [CrossRef]
5. Sharma, H.B.; Vanapalli, K.R.; Cheela, V.S.; Ranjan, V.P.; Jaglan, A.K.; Dubey, B.; Bhattacharya, J. Challenges, opportunities, and innovations for effective solid waste management during and post COVID-19 pandemic. *Resour. Conserv. Recycl.* **2020**, *162*, 105052. [CrossRef]
6. Anh, N.T.M.; Hui, L.; Khoa, V.D.; Mehmood, S. Relational capital and supply chain collaboration for radical and incremental innovation. *Asia Pac. J. Mark. Logist.* **2019**, *31*, 1076–1094.
7. Zhang, A.; Zhong, R.Y.; Farooque, M.; Kang, K.; Venkatesh, V.G. Blockchain-based life cycle assessment: An implementation framework and system architecture. *Resour. Conserv. Recycl.* **2020**, *152*, 104512. [CrossRef]

sustainability MDPI

Article

Assessing Effectiveness of Humanitarian Activities against COVID-19 Disruption: The Role of Blockchain-Enabled Digital Humanitarian Network (BT-DHN)

Sudhanshu Joshi [1,2], Manu Sharma [3,4], Rashmi Prava Das [5], Kamalakanta Muduli [6,7,*], Rakesh Raut [8], B. E. Narkhede [9], Himanshu Shee [10] and Abhishek Misra [3]

1 Operations and Supply Chain Management Area, School of Management, Doon University, Dehradun 248001, India; sudhanshujoshi@doonuniversity.ac.in
2 Faculty of Engineering & Information Technology, University of Technology Sydney, Sidney 2007, Australia
3 Department of Management Studies, Graphic Era Deemed to be University, Dehradun 248002, India; sharmamanu53@gmail.com (M.S.); abhishekmisra@geu.ac.in (A.M.)
4 Guildhall School of Business and Law, London Metropolitan University, London EC2M 6SQ, UK
5 Computer Science Department, CV Raman Global University, Bhubaneswar 752054, India; 2101070068@cgu-odisha.ac.in
6 Department of Mechanical Engineering, Papua New Guinea University of Technology, Lae 411, Papua New Guinea
7 Department of Mechanical Engineering, CV Raman Global University, Bhubaneswar 752054, India
8 Operations & Supply Chain Management, National Institute of Industrial Engineering (NITIE), Mumbai 400087, India; rraut@nitie.ac.in
9 Industrial Engineering & Manufacturing Systems, National Institute of Industrial Engineering (NITIE), Mumbai 400087, India; benarkhede@nitie.ac.in
10 Supply Chain & Logistics Management, College of Business, Victoria University, Melbourne 8001, Australia; Himanshu.Shee@vu.edu.au
* Correspondence: kamalakantam@gmail.com; Tel.: +675-74272286

Citation: Joshi, S.; Sharma, M.; Das, R.P.; Muduli, K.; Raut, R.; Narkhede, B.E.; Shee, H.; Misra, A. Assessing Effectiveness of Humanitarian Activities against COVID-19 Disruption: The Role of Blockchain-Enabled Digital Humanitarian Network (BT-DHN). *Sustainability* **2022**, *14*, 1904. https://doi.org/10.3390/su14031904

Academic Editor: Roberto Cerchione

Received: 15 December 2021
Accepted: 25 January 2022
Published: 7 February 2022

Publisher's Note: MDPI stays neutral with regard to jurisdictional claims in published maps and institutional affiliations.

Copyright: © 2022 by the authors. Licensee MDPI, Basel, Switzerland. This article is an open access article distributed under the terms and conditions of the Creative Commons Attribution (CC BY) license (https://creativecommons.org/licenses/by/4.0/).

Abstract: The COVID-19 pandemic has affected more than 214 countries across the world, disrupting the supply of essential commodities. As the pandemic has spread, humanitarian activities (HAs) have attempted to manage the various situation but appear ineffective due to lack of collaboration and information sharing, inability to respond towards disruption, etc. This study aims to determine and provide insights into the critical factors that may enhance the effectiveness of HAs during the pandemic. A systematic literature review was undertaken to explore critical factors and validated by experts using the fuzzy–Delphi method. These were further assessed to identify the cause-and-effect relationship by means of the fuzzy decision-making trial and laboratory (DEMATEL) method. The results show that building a blockchain-enabled digital humanitarian network (BT-DHN) is the most significant factor during the pandemic. The use of digital platforms for sharing real-time information enhances the effectiveness of HAs. This study offers stakeholders, policymakers, and decision-makers the opportunity to consider these factors in strategic planning to deal with pandemic disruption.

Keywords: humanitarian activities (HAs); humanitarian organization (HO); pandemic disruption; COVID-19; blockchain-enabled digital humanitarian networks (BT-DHN)

1. Introduction

Natural disasters such as earthquakes, tornados, wildfires, floods, etc., inevitably disrupt supply chains regionally or globally [1]. The disruptions are seen in any form: it could be the shortage of materials, a temporary peak in demand of essential items that stimulates fear of resource scarcity, an uncontrollable environment, or many other such undesirable events. Humanitarian supply chains (HSCs) appear to hastily manage such disruptions and uncertainties [2,3]. However, developing an HSC is often more complex when compared to the general commercial supply chain [4]. The disruption caused by the virus outbreaks such as coronavirus (2019-nCoV) in China, the Zika virus, avian influenza A (H7N9), and Ebola virus (Zaire strain) in West Africa created a threat to human health and safety that questions the readiness/preparedness of any organization in meeting such emergency. The rise of the supply of 'essential items' (items of daily needs) and medical equipment (personal protection equipment, surgical masks, and ventilators) faces unprecedented demand for a much higher volume in comparison to the pre-COVID-19 situation [5,6]. The imbalance of demand and supply and the threat to human lives warrants humanitarian activities that offer long-term and short-term aid to the affected population. We define humanitarian activities (HAs) as the humanitarian emergency support offered to rescue any vulnerable individual or a group of individuals in a community by a collaborative effort of humanitarian organizations and their stakeholders. In an emergency, organizations need enhanced operational efficiencies and effective logistics services for vulnerable communities. These organizations, henceforth called humanitarian organizations (HOs), are required to be agile and adaptive to manage the emergency [7,8]. The role of digital technologies, including blockchain, in humanitarian activities is highly significant during a time of emergency [9,10]. BT are useful in the designing and development of the digital humanitarian network. Thus, the BT-enabled DHNs can bring more clarity and accessibility to actors and flawless movement of disaster aids and information across the supply chains [11–13]. Humanitarian aid usually has a linear flow of supplies to the affected areas, especially to regions where the need is higher [14]. During COVID-19, the commercial supply chains deliver the needed supplies. However, humanitarian aids require a vast network and resource prediction until it is needed [15–18]. This acts as a limiting factor for HOs as multiple stakeholders are present in the supply chain. The development of humanitarian strategies and continuous assessment of humanitarian abilities of the cross-sector partners is important for sourcing essentials and strategic supplies [19,20]. The supply chain disruptions can be mitigated using a few operational strategies, including maintaining safety stock or exclusive supplies of healthcare products such as masks, hand sanitizers, protective gear, and ventilators from alternative sources through mobilization of resources [21]. Based on experiences from the past, humanitarian activities should include initiating the action plan and its implementation in cost-effective ways to ensure the flow of goods and services to a vulnerable group of people [22,23]. Therefore, creating a responsive portfolio of customized humanitarian services has become a major concern and topic of discussion by global disaster planners, humanitarian partners, researchers, and practitioners, including the World Health Organization (WHO). Since the 1990s, the WHO has highlighted the need for sustainable partnerships among various stakeholders (including governments, researchers, nonprofit organizations, private firms, and R&D entities), contributing to a variety of HAs in response to disaster mitigation [24,25]. The COVID-19 pandemic is considered the worst crisis since the Second World War [26,27]. As defined by the International Federation of Red Cross and Red Crescent Societies, COVID-19 is categorized as a natural hazard [28]. Disaster risk management has a relationship with the type of disaster, vulnerability, and exposure, as explained in this formula: risk= disaster*vulnerability*exposure [29,30]. For reducing risks, besides disaster prevention, it is required to plan and reduce vulnerability and exposure. Thus, the operational effectiveness in the pandemic situation cannot be seen as a whole; it needs to be broken down into meaningful and efficient sub-systems to measure its effectiveness [31–33]. However, research in this space is quite limited. Table 1 demonstrate the searching pattern from the previous research, the results shows most studies are done using single success factors. However, validation of those success factors using the fuzzy–Delphi method and subsequently assessing through cause-and-effect relationship using the fuzzy decision-making trial and laboratory (DEMATEL) method is new in this study. The present research, therefore, aims to evaluate the HAs in the context of a pandemic situation and to identify these critical factors for their efficiencies and effectiveness. The following research questions are developed to answer this objective.

RQ1. What critical factors contribute to the development of effective HAs in COVID-19?
RQ2. What interrelationship and hierarchy exist between these critical factors (CFs)?

RQ3. To what extent do these critical factors have cause-and-effect interrelationships?

The outcomes of the study will facilitate the disaster planners and strategists to guide their humanitarian supply chain to effectively implement HAs during the pandemic. The study contributes a set of HAs in context to the COVID-19 pandemic. Methodologically, this study employed a systematic literature review followed by the assessment of factors using fuzzy–DEMATEL. The paper is organized as shown below. Section 2 captures the various critical factors based on a systematic literature review. Section 3 describes the research methodology undertaken in the study. Section 4 gives detailed elaboration of the fuzzy–Delphi and fuzzy–DEMATEL methods. Section 4 elaborates the application of methods to validation and cause-and-effect interrelationships computations. Section 5 presents the discussion of the findings of the study. Section 6 highlights the implications, followed by the conclusion and limitations in Section 7.

Table 1. Search criteria.

Search Terms	Initial Search	First Screening	Second Screening	Third Screening	Fourth Screening
"Humanitarian" AND "Pandemic"	15	11	9	8	5
"Humanitarian operations" AND "Pandemic"	21	12	11	10	6
Humanitarian Logistics" AND "COVID-19"	25	20	18	15	12
Critical Success Factors" AND "Humanitarian"	27	11	10	9	5
		Total articles			28

2. Literature

A systematic literature review was undertaken to search articles published from 2000 to 2020. Table 1 presents the search criteria used in the literature review.

The first search resulted in 88 articles. After removing the duplicates, it came down to 54; narrowing down only to journal articles resulted in 48 articles, exclusion of unrelated articles retained 42 articles, and finally, abstract checking resulted in 28 papers. From the selected papers, factors were identified. This followed an expert survey where each expert thoroughly read the description of these critical factors in the questionnaire and evaluated them according to their significance in the enhancement of organizational effectiveness. The detailed elaboration of the factors of HAs to enhance operational activities during a pandemic is discussed in the following section.

Humanitarian Activities (HAs) in Enhancing Operational Effectiveness during the Pandemic

Developing a sustainable humanitarian supply chain (HSC) for managing disasters/emergencies can be viewed as an extension of the traditional supply chain [34,35]. Thus, sustainable HSCs have evolved as a specialized discipline with a focus on social sustainability [36,37]. Various parties (including NGOs, local and regional relief organizations, government agencies, HOs, and beneficiaries) and other stakeholders from the corporate sector comprise a centralized or a decentralized HSC structure [38,39] that aims to relieve the masses at risk. Otherwise, a single actor individually may not have sufficient resources to respond effectively to major disasters, including COVID-19 [40–42]. HAs play a critical role in a disaster. Coordination among humanitarian parties/actors can strengthen and enhance the outcomes through resource and information sharing, decision-making, and conducting joint-field surveys or cluster-based services towards social needs [43–45].

Figure 1 illustrates the conceptual framework on critical factors of HAs influencing operational effectiveness of HOs. These HAs improved resilience through vertical and horizontal coordination among the actors [46–49]. In the light of blockchain technology, the effectiveness of HSC results in a smooth flow of suppliers, information, and resources to the beneficiaries and can be measured in terms of response time by using the common elements of supply chain philosophy: "delivery of the right goods, at the right time, to the right place, and to the right set of people" [50]. Thus, a blockchain-driven HSC can be simply defined as a traceable system available to all stakeholders of HSC for effective roles and responsibility of the disaster migration and effective humanitarian activities [51–53].

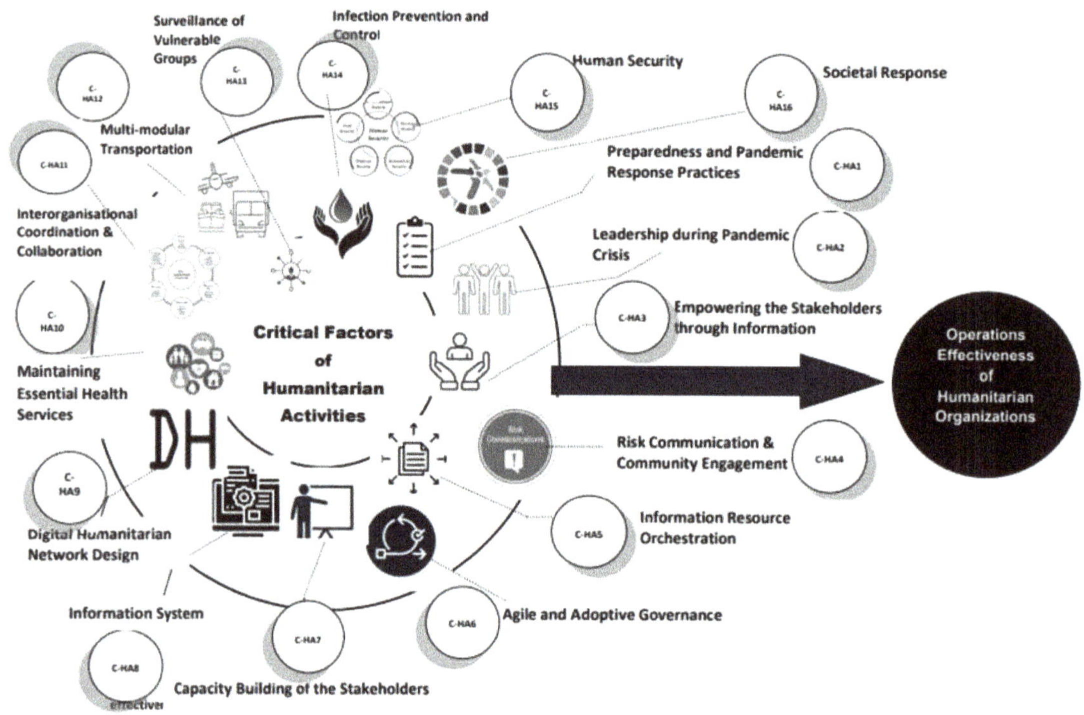

Figure 1. Conceptual framework of critical factors of humanitarian activities influencing humanitarian operations (Source: Authors).

The HAs also result in the development of local and regional infrastructure. Hence, a successful HSC management through HAs thrives to achieve a supply of "essential items" and help in mass evacuation of the community affected by disaster [54,55] through a process of cost-effective flow and storage of goods and materials from the point of origin to the point of consumption for the purpose of meeting the end beneficiary's requirements [56–58]. A typical design of an HSC should be able to manage the available resources efficiently and enable the community to make the right decision by involving local authority through decentralized decision making [59,60]. Usage of technology can help HOs to plan capacity, engage resources, and improve demand prediction. The performance of HSC can be measured by its delivery performance (time, coverage, supply chain responsiveness, and cost involved) [61–63]. The COVID-19 is a global outbreak that leads to a sharp and radical shortage of essential supplies (i.e., PPEs, ventilators, protection masks, sanitizers, and hydroxychloroquine) [64–66]. The HSC partners mean to mitigate the global COVID-19 pandemic situation and to ensure critical supplies to aid recipients. An HSC ensures 'line of sight' along with COVID-19 mitigation, prioritized within the wider set of Has [67,68].

With the increasing pressure due to the loss of human lives, it is necessary to conduct a study that aims to determine the critical factors of Has [69,70]. Multiple stakeholders (parties including the government and private sector) strategically coordinate with each other to perform varieties of HAs to aid recipients [71,72]. Thus, a strategic tie-up has a positive influence on the performance of HSC and increases its sharing capabilities [73–75]. Past literature stressed the feedback mechanism among the stakeholders in a HSCs system for developing a reference model [76–79]. The coordination among humanitarian actors can be increased by cost-effective usage of resources and the involvement of top-level managers in distribution roles [80,81]. Regular interactions between humanitarian actors are essential for the effectiveness of HAs. Effective communication measures to reduce pressure among supply chain actors and optimize the supply of essentials. Usage of ICT ensures the transparency and flawless exchange of information across the HSCs. Additionally, it increases the flexibility, agility, and alignment in emergency decisions. The commitment of humanitarian actors supports the aims of HOs in developing mutual consent towards operational decisions [82]. Effective training of the actors about a pandemic situation helps build capacity to respond more effectively during various disaster situations [83–85]. Various critical success factors are elaborated in Table 2.

Table 2. Critical success factors to enhance operational effectiveness of humanitarian activities.

Critical Factors	Operational Effectiveness during the Pandemic	References
Multi-modal transportation (C-HA1)	Usage of multi-modal transportation can connect all supply nodes, affected areas, and logistics operational areas.	[54]
Leadership during pandemic crisis (C-HA2)	Communicating with teams, stakeholders, and communities during COVID-19 enhances transparency, demonstrates vulnerability, and builds resilience among humanitarian organizations.	[56]
Empowering the stakeholders (C-HA3)	Empowerment of the stakeholders helps the humanitarian organizations to identify clear vision, competency, and coordination across all levels.	[29,38]
Risk communication and community engagement (C-HA4)	Risk communication across stakeholders brings transparency and pro-activeness towards the pandemic situation.	[56]
Information resource orchestration (C-HA5)	Adoption of information resource activities and information behavior activities can meet the need of humanitarian operations.	[49,64]
Agile and adaptive governance (C-HA6)	Participation collaboration and governance become more agile and adaptive during the pandemic.	[60,61]
Information system (C-HA7)	Information system planning should address challenges, value generation processes, and resource base in an effort to improve organizational performance	[63,65,86,87]
Capacity building of stakeholders (C-HA8)	A competency-based teaching approach can improve the intercultural pandemic training among the stakeholders who can further improve interdisciplinary integration, enhancing the overall operational effectiveness.	[57]
Blockchain-enabled digital humanitarian network (BT-DHN) (C-HA9)	Blockchain-enabled digital humanitarian network (BT-DHN) ensures participative management and real-time information flow that uses big data for the humanitarian response for effective relief operations.	[2,4]

Table 2. *Cont.*

Critical Factors	Operational Effectiveness during the Pandemic	References
Maintaining essential health services (C-HA10)	Adjust governance and coordination mechanisms to support timely action for essential health services and adapt to changing contexts and needs.	[20,26,52]
Inter-organizational coordination and collaboration (C-HA11)	Collaborative planning for responding the pandemic (through cooperation, interaction, and collaboration among relief agencies).	[29,38]
Preparedness and pandemic response practices (C-HA12)	Preparedness planning and COVID-19 response practices emerged as the key humanitarian activity among humanitarian actors.	[42,46]
Surveillance for vulnerable groups (C-HA13)	It aims to limit the spread of the pandemic in vulnerable groups (children, women, and the old-age population) by rapid detection, isolation, testing, and management.	[88,89]
Prevention and control (C-HA14)	Infection prevention and control (IPC) is the key humanitarian activity. IPC occupies a unique position in the field of patient safety and quality universal health coverage.	[3]
Human security (C-HA15)	It is protecting human life, especially the vulnerable groups, by involving local government and partners to increase operational effectiveness.	[39]
Societal response (C-HA16)	It is the collective efforts of humanitarian organizations, the corporate world, government, and the community to fight collectively against the pandemic. Based on the principle of 'Respond, Recover and Rebuild', the societal response to the COVID-19 pandemic is a continuous improvement process.	[39,40]

3. Research Methodology

In the past literature, quantitative methods used were either probabilistic techniques, statistics, or both. However, they have several limitations that deal with vagueness and issues of scalability. To delimit these issues, the present study has used an applicable and advanced methodology to assess the effectiveness of the humanitarian activities and to simplify their role during COVID-19 disaster management [85–88]. A three-phase study was conducted, as illustrated in Figure 2. During the first phase, the systematic literature review was conducted to identify HAs, followed by the experts' brainstorming session [41,75]. The detail of experts is presented in Section 4. Based on the responses collected from the experts, validation of the HAs was performed using fuzzy–Delphi. In the second phase, the HAs were assessed using the fuzzy–DEMATEL method to establish the cause-and-effect relationship among them.

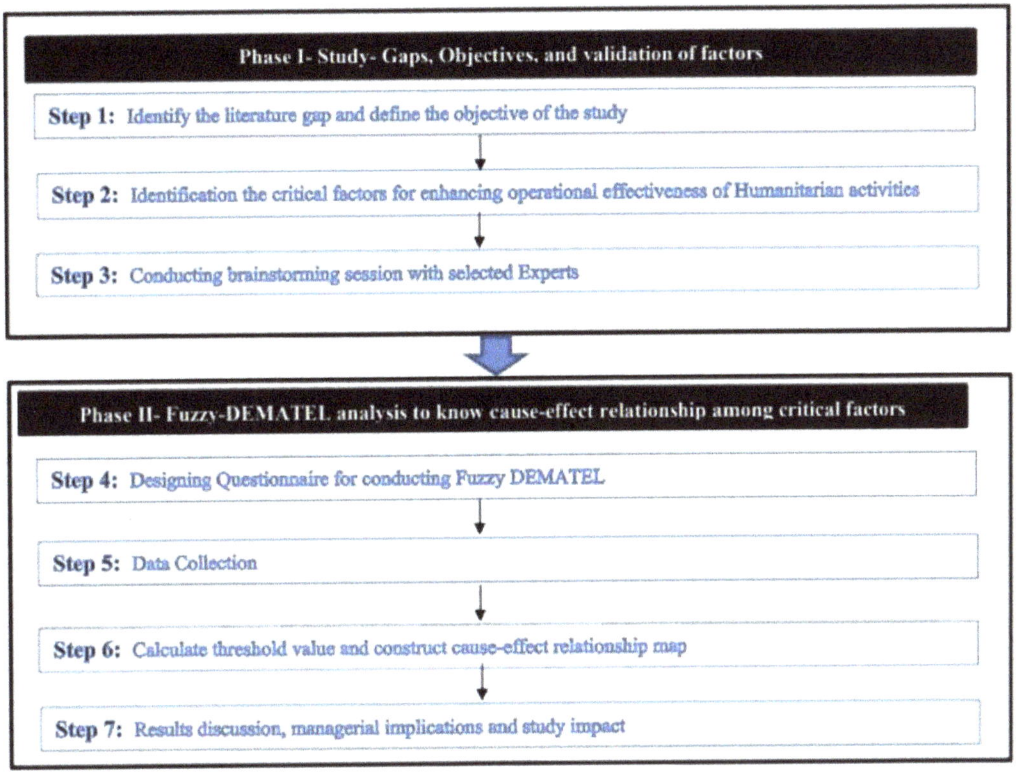

Figure 2. Proposed research framework.

The fuzzy–Delphi and fuzzy–DEMATEL methods are elaborated in the subsequent sub-sections.

3.1. Fuzzy Set Theory

The decision making in the context of HAs is complex due to the involvement of multiple actors as well as the subjectivity in judgment due to ambiguity in the data and information. Thus, fuzzy theory helps the decision-makers to clarify human responses in the crisp form under imprecise and uncertain situations [89,90]. In a fuzzy set, binary numbers 0 and 1 represent each number in an interval [0, 1]. The fuzzy-based analysis can be defined as if 'X' explains a set of elements and the general component of 'X' is explained through 'x' with values $(x_1, x_2, x_3 \ldots\ldots\ldots x_n)$. The fuzzy set C for X can be stated as $\{(x,\ \mu_C(x)) \mid x \in X\}$. The membership of this fuzzy set C can be defined through $\mu_C(x)$.

Let us assume, 'A' and 'B' are two TFNs and represented as A = (p_1, q_1, r_1) and B = (p_2, q_2, r_2). The membership function for the TFN (p, q, r) is calculated using the expression provided in Equation (1).

$$\mu_C(x) = \left\{ \begin{array}{c} 0, x \le p \\ \frac{x-p}{q-p}, x \in [p,q] \\ \frac{x-r}{q-r}, x \in [q,r] \\ 0, \text{otherwise} \end{array} \right\} \quad (1)$$

Then, the algebraic operations for A and B as per the extension principle:

1. A⊕B: $(p_1, q_1, r_1) \oplus (p_2, q_2, r_2) = (p_1 + p_2, q_1 + q_2, r_1 + r_2)$
2. A⊖B: $(p_1, q_1, r_1) \ominus (p_2, q_2, r_2) = (p_1 - p_2, q_1 - q_2, r_1 - r_2)$

3. A⊗B: $(p_1, q_1, r_1) \otimes (p_2, q_2, r_2) \cong (p_1 p_2, q_1 q_2, r_1 r_2)$
4. λ(A⊗B): $y \otimes (p_1, q_1, r_1) = (yp_1, yq_1, yr_1)$
5. A⊘B: $(p_1, q_1, r_1) \oslash (p_2, q_2, r_2) \cong (p_1/r_2, q_1/q_2, r_1/p_2)$

3.2. *Fuzzy–Delphi Method*

The fuzzy-based Delphi [78] has the capability to capture vagueness in data. Several studies have used this method for measuring firm performance [90], performance of green supply chain management [91], technology selection [92], and logistics [90–92]. This study has applied fuzzy–Delphi to obtain the joint decision making that aims to assess the critical factors for HAs to develop humanitarian supply chains. The process is elaborated in the following steps.

Step 1: It includes the extraction of HAs from the existing literature. The extraction is exhibited in Figure 1.

Step 2: The identified HAs were shared with the experts. With the help of the linguistic scale (Table 3), the HAs are evaluated. Assuming fuzzy number $\widetilde{z}_{ij}$ to be the jth evaluation of barriers of the ith expert of n experts,

$$\begin{gathered}\widetilde{z}_{ij} = (a_{ij}, b_{ij}, \, c_{ij}) \\ \text{for } i = 1,\, 2,\, 3, \ldots, n \text{ and } j = 1,\, 2,\, 3, \ldots, m\end{gathered} \tag{2}$$

Then, the fuzzy weights of barriers $\widetilde{a}_j$ are given as follows: $\widetilde{a}_j = (a_j,\, b_j,\, c_j)$, where:

$$\begin{gathered}a_j = \min\,(a_{ij}), \\ b_j = \left(\prod_{i=1}^{n}(b_{ij})\right)^{1/n} \\ c_j = \max\,(c_{ij}), \text{where}, i = 1,\, 2, \ldots, n, j = 1,\, 2, \ldots m\end{gathered} \tag{3}$$

Step 3: This final step uses mean method Sj through Equation (4).

$$Sj = \frac{(a_j + b_j + c_j)}{3},\; j = 1,2,3, \ldots m \tag{4}$$

The evaluation of critical factors is based on the following condition:

a) Acceptance of factor: When the value of S_j is greater or equal to the threshold value (α)

b) Rejection of the factor: When the value of S_j is less than a threshold value (α)

Table 3. Scale labeling.

Terms for Scale	Number	Linguistic Terms
Very influential (VI)	4	(0.75, 1.0, 1.0)
High influence (HI)	3	(0.5, 0.75, 1.0)
Low influence (LI)	2	(0.25, 0.5, 0.75)
Very low influence (VLI)	1	(0, 0.25, 0.5)
No influence (NI)	0	(0, 0, 0.25)

3.3. *Fuzzy DEMATEL*

In a multi-variable decision making fuzzy and complex supply chain management problem, fuzzy–DEMATEL can be used as an effective tool [10,51,69,85]. Broadly, the mathematical process can be explained as follows:

Step 1: Goal setting and criteria identification

Step 2: Factors identification to evaluate effect between factors using pairwise comparison.

Step 3: Define the fuzzy linguistic scale. Table 3 explains the linguistic terms used in the study.

Step 4: Development of fuzzy direct-relation matrix Zk. Zk = [Zkij] where Z is a n × n non-negative matrix; Zij represents the direct impact of factor i on factor j, and, when i = j, the diagonal elements Zij = 0.

Step 5: Establishment of the cause-and-effect model: Compute the total-relation matrix T using the formula in Equation (5), where n × n identity matrix is represented with I. Upper, and lower values are calculated separately

$$T = D(I - D)^{-1} \tag{5}$$

Step 6: The cause-and-effect group factors provides the visualization of the complex interrelationships among factors and are highly significant for decision-makers.

4. Research Framework

The methods are applied sequentially as shown in Figure 2. The framework is elaborated as follows:

4.1. Phase 1: Identification and Validation of Critical Factors for HAs through Brainstorming

From the literature review, sixteen critical success factors related to HAs were identified. A brainstorming session was conducted online to identify the perception of health officials and humanitarian organizations (NGOs and private healthcare staff). The data were collected through a questionnaire with an additional sheet to include any extra critical factors. A panel of 11 experts with different expertise over 10 years were engaged in the brainstorming session. The details of the experts are given in Table 4.

Table 4. Details of experts.

Expert Code	Designation	Age (Years)	Industry	Experience (Years)	Expertise
E1	Healthcare professional	>45	Health care	>15	Patient care
E2	Healthcare professional	>45	Health care	>15	Patient care
E3	Disaster management expert	>35	Healthcare	>12	Healthcare
E4	Disaster management expert	>40	Healthcare	>15	Healthcare
E5	Disaster management expert	>40	Healthcare	>15	Healthcare
E6	NGO	>40	Social well being	>15	Societal issue
E7	Manager	>35	Healthcare	>15	Healthcare
E8	Healthcare staff	>35	Healthcare	>10	Patient care
E9	Professor	>45	Higher education	>20	Healthcare
E10	Professor	>45	Higher education	>20	Healthcare
E11	Healthcare staff	>35	Healthcare	>10	Patient care

The responses were collected from the experts based on the linguistic label shown in Table 3. A threshold value was set more than 0.60 for exclusion and inclusion of the factors based on the previous literature. The experts were also asked to include any factor that they feel can influence the HAs during the pandemic. However, the experts did not suggest any change and were satisfied with the factors they were provided. Through the fuzzy–Delphi method, the factors were assessed and validated. Section 3.1 discussed the steps for computing Sj, and its final values are exhibited in Table 5.

Table 5. Scores for variables were undertaken using fuzzy–Delphi.

S. N	Critical Factors for HAs	l	m	u	S
1	Multi-modal transportation (C-HA1)	0.25	0.89	1.00	0.712
2	Leadership during pandemic crisis (C-HA2)	0.25	0.80	1.00	0.682
3	Empowering the stakeholders through information (C-HA3)	0.25	0.84	1.00	0.697
4	Risk communication and community engagement (C-HA4)	0.25	0.82	1.00	0.689
5	Information resource orchestration (C-HA5)	0.30	0.82	1.00	0.706
6	Agile and adaptive governance (C-HA6)	0.25	0.75	1.00	0.667
7	Information system (C-HA7)	0.25	0.84	1.00	0.697
8	Capacity building of stakeholders (C-HA8)	0.25	0.86	1.00	0.705
9	Prevention and control (C-HA9)	0.25	0.82	1.00	0.689
10	Maintaining essential health services (C-HA10)	0.25	0.80	1.00	0.682
11	Inter-organizational coordination and collaboration (C-HA11)	0.25	0.75	1.00	0.667
12	Preparedness and pandemic response practices (C-HA12)	0.25	0.80	1.00	0.682
13	Surveillance for vulnerable groups (C-HA13)	0.25	0.82	1.00	0.689
14	Blockchain-enabled digital humanitarian network (BT-DHN) design (C-HA14)	0.25	0.77	1.00	0.673
15	Human security (C-HA15)	0.25	0.82	1.00	0.689
16	Societal response (C-HA16)	0.00	0.70	1.00	0.568

The values of Sj in Table 5 suggest that all the variables identified from the literature are valid and must be undertaken for the study as all the values are higher than 0.60.

4.2. *Fuzzy–DEMATEL for Cause-and-Effect Analysis*

The fuzzy–DEMATEL was applied to establish a cause-and-effect relationship among the sixteen critical factors. The factors were assessed on a linguistic scale mentioned in Table 3. The normalized fuzzy numbers and total relation matrix derived from the step-by-step process are shown in Table 6.

Table 6. Total normalized direct-relation matrix (X) for *l*, *m*, and *u*.

								(*l*)								
Factors	C-HA1	C-HA2	C-HA3	C-HA4	C-HA5	C-HA6	C-HA7	C-HA8	C-HA9	C-HA 10	C-HA11	C-HA12	C-HA13	C-HA14	C-HA15	C-HA16
C-HA1	0	0.0162	0.0129	0.0323	0.0356	0.0209	0.0210	0.0339	0.0387	0.0242	0.0355	0.0355	0.0338	0.0388	0.0355	0.0258
C-HA2	0.0209	0	0.0178	0.0501	0.0162	0.0161	0.0193	0.0388	0.0242	0.0194	0.0258	0.0145	0.0210	0.0194	0.0226	0.0340
C-HA3	0.0210	0.0194	0	0.0355	0.0323	0.0097	0.0000	0.0388	0.0291	0.0243	0.0323	0.0242	0.0259	0.0275	0.0064	0.0161
C-HA4	0.0308	0.0194	0.0000	0	0.0533	0.0178	0.0210	0.0323	0.0226	0.0194	0.0242	0.0178	0.0162	0.0178	0.0226	0.0356
C-HA5	0.0370	0.0032	0.0323	0.0000	0	0.0355	0.0178	0.0323	0.0146	0.0323	0.0388	0.0291	0.0275	0.0290	0.0194	0.0356
C-HA6	0.0306	0.0243	0.0032	0.0242	0.0355	0	0.0178	0.0355	0.0178	0.0355	0.0307	0.0323	0.0355	0.0211	0.0339	0.0242
C-HA7	0.0322	0.0194	0.0242	0.0210	0.0178	0.0161	0	0.0291	0.0178	0.0355	0.0501	0.0323	0.0469	0.0356	0.0371	0.0436
C-HA8	0.0258	0.0226	0.0274	0.0162	0.0178	0.0178	0.0355	0	0.0064	0.0064	0.0112	0.0291	0.0469	0.0339	0.0371	0.0372
C-HA9	0.0274	0.0177	0.0307	0.0177	0.0194	0.0226	0.0355	0.0355	0	0.0194	0.0340	0.0323	0.0323	0.0501	0.0371	0.0372
C-HA10	0.0322	0.0193	0.0259	0.0162	0.0113	0.0209	0.0161	0.0178	0.0178	0	0.0533	0.0194	0.0178	0.0226	0.0355	0.0340
C-HA11	0.0193	0.0323	0.0355	0.0097	0.0259	0.0290	0.0355	0.0290	0.0533	0.0178	0	0.0517	0.0501	0.0210	0.0517	0.0355
C-HA12	0.0291	0.0307	0.0355	0.0259	0.0323	0.0419	0.0484	0.0484	0.0533	0.0178	0.0533	0	0.0533	0.0226	0.0178	0.0404
C-HA13	0.0323	0.0371	0.0404	0.0226	0.0355	0.0209	0.0322	0.0323	0.0355	0.0178	0.0355	0.0355	0	0.0194	0.0194	0.0356
C-HA14	0.0436	0.0485	0.0420	0.0161	0.0388	0.0210	0.0355	0.0469	0.0355	0.0178	0.0355	0.0178	0.0355	0	0.0517	0.0355
C-HA15	0.0420	0.0452	0.0452	0.0161	0.0259	0.0242	0.0355	0.0533	0.0355	0.0178	0.0242	0.0178	0.0355	0.0178	0	0.0501
C-HA16	0.0420	0.0307	0.0501	0.0420	0.0194	0.0128	0.0404	0.0371	0.0517	0.0371	0.0436	0.0210	0.0355	0.0211	0.0178	0
								(*m*)								
C-HA1	0	0.0340	0.0307	0.0501	0.0534	0.0387	0.0387	0.0517	0.0565	0.0420	0.0532	0.0533	0.0516	0.0565	0.0533	0.0436
C-HA2	0.0387	0	0.0355	0.0679	0.0242	0.0339	0.0371	0.0565	0.0420	0.0372	0.0436	0.0323	0.0388	0.0372	0.0404	0.0517
C-HA3	0.0387	0.0372	0	0.0533	0.0501	0.0274	0.0178	0.0566	0.0469	0.0420	0.0501	0.0419	0.0436	0.0453	0.0242	0.0339
C-HA4	0.0486	0.0372	0.0178	0	0.0711	0.0355	0.0388	0.0501	0.0404	0.0372	0.0420	0.0356	0.0340	0.0355	0.0404	0.0533
C-HA5	0.0548	0.0210	0.0501	0.0178	0	0.0533	0.0355	0.0501	0.0324	0.0501	0.0565	0.0468	0.0452	0.0468	0.0371	0.0533
C-HA6	0.0484	0.0420	0.0210	0.0420	0.0533	0	0.0355	0.0533	0.0355	0.0533	0.0485	0.0501	0.0533	0.0389	0.0517	0.0420
C-HA7	0.0500	0.0372	0.0420	0.0387	0.0355	0.0339	0	0.0469	0.0355	0.0533	0.0679	0.0501	0.0647	0.0533	0.0549	0.0614
C-HA8	0.0435	0.0404	0.0452	0.0340	0.0355	0.0355	0.0533	0	0.0242	0.0242	0.0290	0.0469	0.0647	0.0517	0.0549	0.0550
C-HA9	0.0452	0.0355	0.0484	0.0354	0.0372	0.0403	0.0533	0.0533	0	0.0372	0.0517	0.0501	0.0501	0.0679	0.0549	0.0549
C-HA10	0.0500	0.0371	0.0436	0.0340	0.0291	0.0386	0.0339	0.0355	0.0355	0	0.0711	0.0372	0.0355	0.0404	0.0533	0.0517
C-HA11	0.0371	0.0500	0.0533	0.0275	0.0437	0.0468	0.0533	0.0468	0.0711	0.0355	0	0.0695	0.0679	0.0388	0.0695	0.0533

Table 6. *Cont.*

								(m)								
C-HA12	0.0468	0.0484	0.0533	0.0436	0.0501	0.0597	0.0661	0.0661	0.0711	0.0355	0.0711	0	0.0711	0.0404	0.0355	0.0582
C-HA13	0.0501	0.0549	0.0581	0.0404	0.0533	0.0387	0.0500	0.0501	0.0533	0.0355	0.0533	0.0533	0	0.0372	0.0371	0.0533
C-HA14	0.0614	0.0663	0.0598	0.0339	0.0566	0.0388	0.0533	0.0647	0.0533	0.0355	0.0533	0.0355	0.0533	0	0.0695	0.0533
C-HA15	0.0597	0.0630	0.0630	0.0339	0.0437	0.0420	0.0533	0.0711	0.0533	0.0355	0.0420	0.0355	0.0533	0.0355	0	0.0678
C-HA16	0.0598	0.0484	0.0679	0.0598	0.0371	0.0306	0.0581	0.0549	0.0695	0.0549	0.0614	0.0387	0.0533	0.0389	0.0355	0
								(u)								
C-HA1	0	0.0485	0.0484	0.0679	0.0630	0.0549	0.0565	0.0678	0.0678	0.0598	0.0646	0.0646	0.0613	0.0630	0.0646	0.0598
C-HA2	0.0533	0	0.0533	0.0711	0.0388	0.0517	0.0533	0.0711	0.0565	0.0549	0.0582	0.0484	0.0566	0.0549	0.0582	0.0614
C-HA3	0.0533	0.0550	0	0.0711	0.0678	0.0420	0.0355	0.0679	0.0598	0.0566	0.0614	0.0549	0.0582	0.0550	0.0420	0.0517
C-HA4	0.0598	0.0550	0.0355	0	0.0711	0.0533	0.0565	0.0679	0.0566	0.0550	0.0565	0.0518	0.0518	0.0533	0.0581	0.0662
C-HA5	0.0629	0.0388	0.0679	0.0355	0	0.0711	0.0533	0.0647	0.0469	0.0679	0.0711	0.0598	0.0598	0.0581	0.0549	0.0662
C-HA6	0.0565	0.0582	0.0388	0.0565	0.0711	0	0.0533	0.0711	0.0533	0.0711	0.0663	0.0662	0.0695	0.0550	0.0695	0.0598
C-HA7	0.0629	0.0533	0.0565	0.0549	0.0533	0.0517	0	0.0647	0.0533	0.0711	0.0695	0.0678	0.0679	0.0695	0.0711	0.0711
C-HA8	0.0597	0.0566	0.0598	0.0485	0.0533	0.0533	0.0711	0	0.0420	0.0420	0.0468	0.0647	0.0679	0.0695	0.0711	0.0663
C-HA9	0.0613	0.0517	0.0598	0.0516	0.0550	0.0549	0.0711	0.0711	0	0.0549	0.0695	0.0679	0.0679	0.0711	0.0711	0.0630
C-HA10	0.0630	0.0533	0.0582	0.0485	0.0452	0.0516	0.0517	0.0533	0.0533	0	0.0711	0.0549	0.0533	0.0582	0.0711	0.0663
C-HA11	0.0501	0.0630	0.0678	0.0453	0.0582	0.0613	0.0711	0.0646	0.0711	0.0533	0	0.0711	0.0711	0.0565	0.0711	0.0646
C-HA12	0.0614	0.0630	0.0678	0.0614	0.0678	0.0645	0.0678	0.0678	0.0711	0.0533	0.0711	0	0.0711	0.0582	0.0533	0.0647
C-HA13	0.0662	0.0694	0.0678	0.0582	0.0678	0.0565	0.0678	0.0679	0.0711	0.0533	0.0711	0.0711	0	0.0549	0.0549	0.0647
C-HA14	0.0711	0.0711	0.0711	0.0517	0.0711	0.0565	0.0711	0.0679	0.0711	0.0533	0.0711	0.0533	0.0711	0	0.0711	0.0678
C-HA15	0.0694	0.0694	0.0711	0.0517	0.0582	0.0581	0.0711	0.0711	0.0711	0.0533	0.0565	0.0533	0.0711	0.0533	0	0.0695
C-HA16	0.0711	0.0662	0.0711	0.0711	0.0549	0.0452	0.0711	0.0711	0.0711	0.0711	0.0678	0.0549	0.0711	0.0550	0.0533	0

Further, the total relation matrix is obtained by using the formula described in Equation (5) and shown in Table 7.

Table 7. Total relation matrix.

								(l)								
Factors	C-HA1	C-HA2	C-HA3	C-HA4	C-HA5	C-HA6	C-HA7	C-HA8	C-HA9	C-HA 10	C-HA11	C-HA12	C-HA13	C-HA14	C-HA15	C-HA16
C-HA1	0.0240	0.0363	0.0356	0.0481	0.0558	0.0378	0.0428	0.0605	0.0612	0.0407	0.0608	0.0557	0.0595	0.0580	0.0576	0.0521
C-HA2	0.0396	0.0160	0.0349	0.0628	0.0334	0.0291	0.0364	0.0594	0.0423	0.0328	0.0459	0.0313	0.0416	0.0354	0.0405	0.0544
C-HA3	0.0385	0.0340	0.0169	0.0475	0.0480	0.0229	0.0171	0.0583	0.0461	0.0364	0.0512	0.0400	0.0453	0.0426	0.0244	0.0361
C-HA4	0.0491	0.0342	0.0184	0.0133	0.0678	0.0313	0.0379	0.0529	0.0404	0.0332	0.0448	0.0345	0.0371	0.0340	0.0403	0.0557

Table 7. *Cont.*

								(l)								
C-HA5	0.0566	0.0213	0.0514	0.0157	0.0183	0.0493	0.0363	0.0552	0.0357	0.0468	0.0611	0.0475	0.0506	0.0461	0.0393	0.0570
C-HA6	0.0514	0.0419	0.0240	0.0391	0.0535	0.0158	0.0375	0.0592	0.0388	0.0500	0.0540	0.0508	0.0585	0.0389	0.0538	0.0481
C-HA7	0.0561	0.0408	0.0479	0.0389	0.0395	0.0335	0.0231	0.0569	0.0433	0.0524	0.0762	0.0538	0.0731	0.0553	0.0600	0.0698
C-HA8	0.0457	0.0399	0.0464	0.0316	0.0357	0.0315	0.0530	0.0239	0.0272	0.0216	0.0340	0.0460	0.0679	0.0498	0.0546	0.0586
C-HA9	0.0512	0.0388	0.0534	0.0354	0.0407	0.0391	0.0570	0.0629	0.0246	0.0367	0.0601	0.0531	0.0591	0.0692	0.0597	0.0631
C-HA10	0.0506	0.0360	0.0439	0.0306	0.0286	0.0342	0.0338	0.0402	0.0381	0.0142	0.0730	0.0370	0.0397	0.0386	0.0537	0.0545
C-HA11	0.0457	0.0543	0.0606	0.0300	0.0486	0.0474	0.0594	0.0599	0.0784	0.0370	0.0302	0.0738	0.0786	0.0439	0.0747	0.0646
C-HA12	0.0569	0.0540	0.0617	0.0469	0.0571	0.0610	0.0732	0.0798	0.0801	0.0391	0.0837	0.0273	0.0844	0.0479	0.0459	0.0714
C-HA13	0.0549	0.0556	0.0615	0.0405	0.0556	0.0376	0.0529	0.0592	0.0584	0.0353	0.0614	0.0562	0.0271	0.0401	0.0417	0.0610
C-HA14	0.0695	0.0699	0.0668	0.0369	0.0615	0.0397	0.0591	0.0773	0.0615	0.0376	0.0645	0.0423	0.0652	0.0241	0.0760	0.0652
C-HA15	0.0658	0.0648	0.0677	0.0363	0.0475	0.0410	0.0573	0.0808	0.0595	0.0364	0.0519	0.0405	0.0630	0.0402	0.0240	0.0763
C-HA16	0.0663	0.0514	0.0724	0.0605	0.0425	0.0310	0.0623	0.0659	0.0758	0.0551	0.0716	0.0447	0.0635	0.0442	0.0431	0.0292
								(m)								
C-HA1	0.1108	0.1327	0.1362	0.1400	0.1529	0.1281	0.1418	0.1709	0.1636	0.1322	0.1696	0.1540	0.1678	0.1548	0.1590	0.1608
C-HA2	0.1334	0.0866	0.1258	0.1455	0.1128	0.1109	0.1262	0.1591	0.1356	0.1158	0.1447	0.1209	0.1400	0.1236	0.1327	0.1525
C-HA3	0.1319	0.1209	0.0906	0.1302	0.1355	0.1047	0.1070	0.1577	0.1388	0.1189	0.1495	0.1290	0.1432	0.1302	0.1166	0.1345
C-HA4	0.1433	0.1221	0.1106	0.0807	0.1558	0.1137	0.1282	0.1536	0.1344	0.1168	0.1444	0.1247	0.1365	0.1228	0.1332	0.1546
C-HA5	0.1541	0.1130	0.1460	0.1037	0.0940	0.1344	0.1302	0.1597	0.1334	0.1332	0.1641	0.1408	0.1533	0.1380	0.1358	0.1597
C-HA6	0.1507	0.1343	0.1210	0.1277	0.1467	0.0859	0.1328	0.1652	0.1380	0.1377	0.1588	0.1453	0.1626	0.1325	0.1513	0.1527
C-HA7	0.1612	0.1389	0.1500	0.1328	0.1389	0.1257	0.1071	0.1695	0.1483	0.1453	0.1866	0.1541	0.1830	0.1540	0.1632	0.1800
C-HA8	0.1425	0.1299	0.1400	0.1179	0.1270	0.1161	0.1451	0.1108	0.1240	0.1077	0.1366	0.1381	0.1687	0.1405	0.1493	0.1599
C-HA9	0.1556	0.1361	0.1544	0.1285	0.1392	0.1303	0.1564	0.1743	0.1118	0.1293	0.1700	0.1524	0.1684	0.1665	0.1619	0.1725
C-HA10	0.1460	0.1250	0.1365	0.1159	0.1190	0.1177	0.1254	0.1427	0.1333	0.0822	0.1730	0.1283	0.1402	0.1285	0.1473	0.1548
C-HA11	0.1551	0.1555	0.1659	0.1276	0.1512	0.1425	0.1633	0.1764	0.1860	0.1338	0.1287	0.1770	0.1923	0.1465	0.1811	0.1789
C-HA12	0.1707	0.1597	0.1716	0.1482	0.1639	0.1598	0.1813	0.2008	0.1924	0.1400	0.2029	0.1191	0.2030	0.1549	0.1580	0.1906
C-HA13	0.1589	0.1520	0.1619	0.1332	0.1532	0.1285	0.1522	0.1704	0.1617	0.1276	0.1710	0.1552	0.1197	0.1382	0.1443	0.1702
C-HA14	0.1797	0.1722	0.1736	0.1358	0.1652	0.1366	0.1647	0.1952	0.1714	0.1359	0.1811	0.1482	0.1812	0.1116	0.1841	0.1814
C-HA15	0.1725	0.1638	0.1709	0.1319	0.1481	0.1345	0.1594	0.1945	0.1657	0.1315	0.1650	0.1430	0.1752	0.1411	0.1128	0.1882
C-HA16	0.1745	0.1523	0.1769	0.1567	0.1450	0.1262	0.1657	0.1818	0.1830	0.1509	0.1857	0.1485	0.1773	0.1465	0.1501	0.1269
								(u)								
C-HA1	0.6146	0.6302	0.6464	0.6282	0.6568	0.6060	0.6709	0.7324	0.6748	0.6366	0.7055	0.6654	0.7030	0.6513	0.6834	0.6966

Table 7. *Cont.*

(*u*)																
C-HA2	0.6183	0.5397	0.6046	0.5888	0.5895	0.5604	0.6209	0.6845	0.6181	0.5880	0.6503	0.6048	0.6495	0.5990	0.6303	0.6493
C-HA3	0.6107	0.5842	0.5470	0.5814	0.6084	0.5456	0.5972	0.6734	0.6134	0.5824	0.6457	0.6032	0.6429	0.5918	0.6080	0.6327
C-HA4	0.6297	0.5964	0.5944	0.5265	0.6233	0.5674	0.6293	0.6874	0.6233	0.5937	0.6549	0.6132	0.6511	0.6028	0.6358	0.6593
C-HA5	0.6520	0.6011	0.6429	0.5798	0.5766	0.6008	0.6458	0.7059	0.6348	0.6240	0.6889	0.6402	0.6792	0.6260	0.6528	0.6796
C-HA6	0.6699	0.6406	0.6397	0.6195	0.6653	0.5558	0.6698	0.7371	0.6635	0.6485	0.7089	0.6687	0.7121	0.6457	0.6896	0.6986
C-HA7	0.6911	0.6513	0.6708	0.6329	0.6646	0.6184	0.6346	0.7482	0.6794	0.6630	0.7280	0.6850	0.7270	0.6734	0.7065	0.7245
C-HA8	0.6468	0.6147	0.6330	0.5893	0.6245	0.5823	0.6591	0.6425	0.6279	0.5980	0.6641	0.6412	0.6834	0.6334	0.6639	0.6769
C-HA9	0.6922	0.6523	0.6763	0.6322	0.6688	0.6237	0.7038	0.7569	0.6312	0.6511	0.7307	0.6879	0.7299	0.6776	0.7092	0.7201
C-HA10	0.6360	0.5990	0.6184	0.5768	0.6039	0.5687	0.6284	0.6784	0.6248	0.5447	0.6716	0.6195	0.6564	0.6103	0.6508	0.6628
C-HA11	0.6801	0.6604	0.6815	0.6251	0.6696	0.6276	0.7016	0.7491	0.6956	0.6480	0.6637	0.6889	0.7307	0.6628	0.7069	0.7193
C-HA12	0.7038	0.6732	0.6945	0.6521	0.6919	0.6430	0.7123	0.7670	0.7091	0.6612	0.7447	0.6362	0.7450	0.6776	0.7053	0.7338
C-HA13	0.7068	0.6776	0.6934	0.6482	0.6904	0.6348	0.7110	0.7657	0.7079	0.6599	0.7433	0.7013	0.6772	0.6736	0.7054	0.7324
C-HA14	0.7293	0.6963	0.7141	0.6588	0.7106	0.6509	0.7320	0.7853	0.7259	0.6769	0.7621	0.7031	0.7626	0.6387	0.7380	0.7540
C-HA15	0.7007	0.6690	0.6872	0.6343	0.6728	0.6275	0.7047	0.7586	0.6987	0.6513	0.7206	0.6764	0.7340	0.6633	0.6440	0.7271
C-HA16	0.7120	0.6755	0.6967	0.6605	0.6795	0.6250	0.7146	0.7692	0.7086	0.6762	0.7411	0.6876	0.7440	0.6744	0.7049	0.6726

The value for the causal diagram is obtained $(D + R)$ and $(D - R)$ and shown in Table 8.

Table 8. Values for the causal diagram.

	Di			Ri			Di + Ri			Di − Ri			Crisp Di + Ri	Crisp Di − Ri
	l	m	u	l	m	u	l	m	u	l	m	u		
C-HA1	0.7865	2.3754	10.6020	0.8219	2.4409	10.6942	1.6084	4.8163	21.2962	−9.9077	−0.0655	9.7801	7.4845	−0.0734
C-HA2	0.6357	2.0659	9.7961	0.6891	2.1950	10.1616	1.3248	4.2609	19.9577	−9.5259	−0.1291	9.1070	6.8647	−0.1523
C-HA3	0.6053	2.0391	9.6681	0.7634	2.3318	10.4412	1.3686	4.3709	20.1093	−9.8359	−0.2927	8.9047	6.9706	−0.3208
C-HA4	0.6249	2.0755	9.8886	0.6143	2.0564	9.8343	1.2393	4.1319	19.7229	−9.2094	0.0191	9.2742	6.7284	0.0038
C-HA5	0.6883	2.1934	10.2305	0.7340	2.2485	10.3966	1.4223	4.4419	20.6271	−9.7083	−0.0551	9.4965	7.1040	−0.0798
C-HA6	0.7154	2.2433	10.6334	0.5821	1.9955	9.6379	1.2975	4.2388	20.2714	−8.9226	0.2477	10.0513	6.8942	0.2898
C-HA7	0.8205	2.4385	10.8986	0.7390	2.2872	10.7361	1.5596	4.7257	21.6347	−9.9155	0.1513	10.1596	7.4653	0.1013
C-HA8	0.6675	2.1542	10.1811	0.9520	2.6826	11.6417	1.6194	4.8368	21.8227	−10.9742	−0.5285	9.2291	7.5758	−0.5656
C-HA9	0.8040	2.4078	10.9438	0.8113	2.4215	10.6369	1.6153	4.8293	21.5807	−9.8329	−0.0137	10.1325	7.5354	0.0175
C-HA10	0.6468	2.1158	9.9505	0.6052	2.0387	10.1036	1.2520	4.1546	20.0542	−9.4569	0.0771	9.3453	6.7966	0.0093
C-HA11	0.8871	2.5618	10.9109	0.9245	2.6316	11.2241	1.8116	5.1934	22.1350	−10.3370	−0.0698	9.9864	7.8864	−0.1082
C-HA12	0.9704	2.7169	11.1506	0.7345	2.2785	10.5225	1.7049	4.9954	21.6731	−9.5521	0.4384	10.4161	7.6732	0.3519
C-HA13	0.7989	2.3982	11.1290	0.9143	2.6124	11.2279	1.7132	5.0107	22.3570	−10.4290	−0.2142	10.2147	7.7816	−0.1691
C-HA14	0.9173	2.6178	11.4387	0.7084	2.2302	10.3018	1.6257	4.8480	21.7405	−9.3845	0.3876	10.7303	7.5725	0.3926

Table 8. *Cont.*

	Di			Ri			Di + Ri			Di − Ri			Crisp Di + Ri	Crisp Di − Ri
	l	m	u	l	m	u	l	m	u	l	m	u		
C-HA15	0.8528	2.4981	10.9703	0.7892	2.3806	10.8348	1.6420	4.8788	21.8050	−9.9820	0.1175	10.1810	7.6048	0.0757
C-HA16	0.8794	2.5481	11.1423	0.9174	2.6183	11.1395	1.7968	5.1664	22.2818	−10.2601	−0.0702	10.2249	7.8870	−0.0632

Based on the $(D - R)$ values, the cause-and-effect relationship is established among the factors. The impact results are shown in Table 9.

Table 9. Impact results of factors.

Factors	D + R	D − R	Impact
C-HA1	7.4845	−0.0734	Effect
C-HA2	6.8647	−0.1523	Effect
C-HA3	6.9706	−0.3208	Effect
C-HA4	6.7284	0.0038	Cause
C-HA5	7.1040	−0.0798	Effect
C-HA6	6.8942	0.2898	Cause
C-HA7	7.4653	0.1013	Cause
C-HA8	7.5758	−0.5656	Effect
C-HA9	7.5354	0.0175	Cause
C-HA10	6.7966	0.0093	Cause
C-HA11	7.8864	−0.1082	Effect
C-HA12	7.6732	0.3519	Cause
C-HA13	7.7816	−0.1691	Effect
C-HA14	7.5725	0.3926	Cause
C-HA15	7.6048	0.0757	Cause
C-HA16	7.8870	−0.0632	Effect

In order to obtain the digraph and to eliminate minor effects, the threshold value (α) is calculated using Equation (6).

$$\alpha = \left(\sum(i=1)^{n} \sum\_(j=1)^{n}[t\_ij\,]\right)/N = 1.9192 \tag{6}$$

A network relationship map (NRM) was established, based on the value of α (1.91). This presented the significance or strength of the relationship, which are shown in the digraph with an arrow (Figure 3). The values that were more than the threshold value of 1.51 are included in the total relation matrix; see Table 8. A network relationship map (NRM) was established.

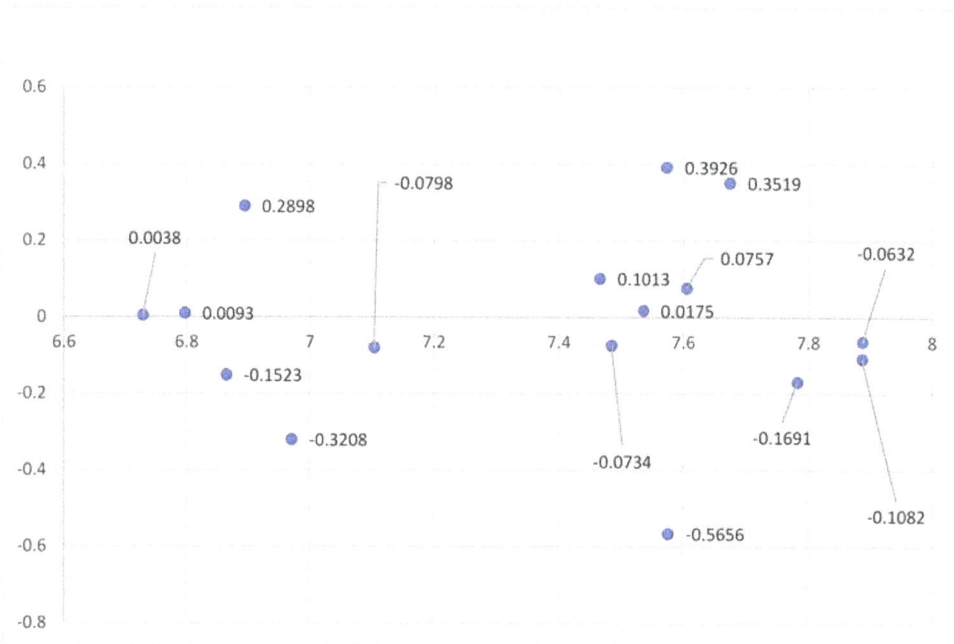

Figure 3. Cause-and-effect relationship.

5. Discussion of Findings

The study explored the key factors that needed to be focused on during pandemics to enhance the operational effectiveness of humanitarian activities (HAs). These critical factors are grouped as causal factors where D−R values are positive, shown in Table 8. The results imply that these causal factors drive the other factors in the system. On the basis of the values of D−R, the factors are categorized into two groups: cause and effect. The causal factors include risk communication and community engagement (C-HA4), agile and adaptive governance (C-HA6), information system (C-HA7), prevention and control (C-HA9), maintaining essential health services (C-HA10), preparedness and pandemic response practices (C-HA12), blockchain-enabled digital humanitarian network (BT-DHN) design (C-HA14), and human security (C-HA15). The causal group factors are elaborated in the following section.

From Table 9, it is visible that blockchain-enabled digital humanitarian network (BT-DHN) design is the most significant factor during the pandemic. Pandemics or disasters are highly complex and develop a challenging environment for humanitarian organizations [19,20]. Intervening during a disaster requires an in-depth understanding of the situation and the context. Social networking sites and social media are used by the people extensively in the front lines of disaster or directly affected to call for help; search for information; and share photos, videos, and text about their personal experiences and communication about safety to their families and friends. People use different digital channels for sharing real-time data to communicate about recent updates [21]. Digital innovation and technologies offer opportunities to save more lives and explore better ways to communicate to meet the needs of affected people during the crisis. Blockchain-enabled digital humanitarian network (BT-DHN) design develops participative management and provides real-time information flow to employ big data for the humanitarian response for effective relief operations. This new method of humanitarian aid is a cost-effective, attractive, and value-neutral way of addressing the needs of those experiencing fragility [11]. This factor regularly encompasses the uses of mobile phones, social media, crisis mapping, crowdsourcing, digital payment systems, and geospatial technologies. The technological innovations have brought the blockchain-enabled digital humanitarian network (BT-DHN) recently to provide support to the people who are the sufferers of a natural disaster or pandemic situation [92] and act as a liaison between the different digital HOs to work on a project. Table 9 shows that the preparedness and pandemic response practices factor (C-HA1) has received the second-highest weightage (0.3519), indicating the importance of this factor in the pandemic situation. Unlike regional events such as hurricanes, earthquakes, or terrorist attacks, a pandemic is a recurring worldwide occurrence with global implications. Pandemic outbreaks highlight the critical significance of effective planning and response to minimize the mortality rate, social and economic disruptions, and organizational risk. The preparedness and pandemic response practices must include the ability to react immediately and faster and be adaptive to the changing scenarios with the changing phases of the pandemic [87,88]. During a pandemic, global supply chains, as well as local supply chains, need to develop and implement planning and response to assess the organizational performance and consider improvements in the light of an event. This factor includes planning, testing, and regular reviews that can enhance the organizational effectiveness of HOs and may place them in a better position to reduce or mitigate the impact of global disruption. It will also provide vigilance, resiliency, and an effective roadmap to direct future activities, which may include an action plan for pandemic planning and response. The third most important factor is agile and adaptive governance (C-HA6), which is required during pandemic times. This is in line with the previous research study on agility in the humanitarian supply chains conducted by Dubey et al. [2], which empirically proved the significance of agility for HSC and HAs. Moreover, the impact of information systems has also been revealed in the study. The current study has a similar direction for managing HSCs that justifies the fourth important causing factor, i.e., information system. The information related to the causes of spread needs to be communicated at a wider level through the stakeholder's participation [90]. The community needs to be empowered with the recent updates, causes, precautions, vaccine (if available), helpline numbers, medical supplies, etc. The pause to the spread can be achieved through this factor. From the results, the factors of multi-modal transportation (C-HA1), leadership during pandemic crisis (C-HA2), empowering the stakeholders (C-HA3), information resource orchestration (C-HA5), capacity building of stakeholders (C-HA8), inter-organizational coordination and collaboration (C-HA11), surveillance for vulnerable groups (C-HA13), and societal response (C-HA16) are categorized as effect group factors.

The previous studies have suggested that effective HSCs are dependent on the people who lead the operations during the pandemic. The role of the leader who initiates and bind the HOs are the game-changer during an emergency situation. The transportation has to be with multiple modes as the essentials, and the healthcare supplies need to be supplied on time, and thus all humanitarian operations and their effectiveness are dependent on transportation and logistics and coordination among the stakeholders such as government, people, NGOs, private organizations, etc.

6. Implications

This paper provides insights for decision-makers, policymakers, and stakeholders to consider the critical factors for implementing strategic actions during COVID-19 pandemic disruption. The increasing engagement of the humanitarian organizations with stakeholders is an extremely positive indicator. The HOs need to work more strategically with other partners, as these may become larger stakeholders in international humanitarian response. The humanitarian system will be more structured, agile, and prepared than it was before. The paper has explored the factors to be considered for developing a 'new normal' environment, which is more prepared for dealing with the pandemic situation. The blockchain-enabled digital humanitarian network (BT-DHN) will act as a base for partnerships and enhance the effectiveness of HAs. The increasing number of technological advancements on the part of humanitarian organizations users offers an opportunity for extending the blockchain-enabled digital humanitarian network (BT-DHN) for detecting physical activity, speech and auditory context, location tracking, etc. The individuals can directly engage in pandemic response activities using a combination of cloud, crowd, and SMS technologies. With the Internet of Things (IoT) technology, the sensor data will match or even outgrow social data soon. This will have a strong impact on the humanitarian efforts. Moreover, satellite imagery can help the delivery of aid in the affected areas. The humanitarian sector needs to connect the data across preparedness, response, and recovery in a pandemic situation. The humanitarian organizations cannot achieve the objectives alone. Thus, collaboration with the private sector is a necessity. The pandemic has created a need for an alliance between the private and public sectors to transform the humanitarian supply chains.

7. Conclusions and Limitations

With the continuous spread of the coronavirus pandemic across the world, disruptions, and falling economies, the catastrophic impact on the crisis-affected population is highly visible. Stretched aid budgets in the humanitarian sector present enormous challenges. The lessons from the COVID-19 have made the organizations prepared for the 'new normal' situation. Mobile technology is aiming to reach seven million people to use life-enhancing mobile-enabled services during disaster preparedness, response, and recovery by 2021. The delivery and impact of assistance by catalyzing partnerships and innovation for new digital humanitarian services advocating for enabling policy environment are to be accelerated. With the help of this paper, we have explored the critical factors to be considered for enhancing the operational effectiveness of humanitarian organizations during the pandemic. This research approach is certainly in line with the increasing trend towards pandemics and new normal situations. The results of this study show blockchain-enabled digital humanitarian network (BT-DHN) (C-HA14) and preparedness and pandemic response practices (C-HA12) are the most critical factors that should be considered to increase the operational effectiveness of HAs during the pandemic. The policymakers and stakeholders will be benefitted by exploring the strength of factors in enhancing the efficiency of HAs to combat the COVID-19 endemic.

This research study has some limitations that are required to be highlighted for future similar studies to consider. The identification and finalization of factors are very challenging. The dynamic environment will develop more factors to be considered for the HOs. Thus, the study has identified sixteen critical factors that may change in future. The study has assessed the factors based on experts from one country, and thus the study may be generalized and replicated to only the developing countries that have a similar condition. The study has investigated the cause-and-effect group developed in the current study that needs to be investigated further with empirical analysis. Furthermore, various perspectives on designing and developing business models for circular economy and their integration with blockchain technology can be extended and empirically developed from the viewpoint of sustainable humanitarian systems.

Author Contributions: Conceptualization, S.J. and M.S.; methodology, S.J. and M.S.; software, M.S. and R.P.D.; formal analysis, M.S. and K.M.; investigation, M.S.; resources, K.M.; data curation, S.J. and M.S.; writing—original draft preparation, S.J., R.R. and B.E.N.; writing—review and editing, S.J., H.S. and A.M.; supervision, K.M. All authors have read and agreed to the published version of the manuscript.

Funding: The research received no external funding.

Institutional Review Board Statement: Not applicable.

Informed Consent Statement: Not applicable.

Conflicts of Interest: The authors declare no conflict of interest.

References

1. Sharma, M.; Luthra, S.; Joshi, S.; Kumar, A. Developiforng a framework for enhancing survivability of sustainable supply chains during and post-COVID-19 pandemic. *Int. J. Logist. Res. Appl.* **2020**, 1–21. [CrossRef]
2. Dubey, R.; Gunasekaran, A.; Bryde, D.J.; Dwivedi, Y.; Papadopoulos, T. Blockchain technology for enhancing swift-trust, collaboration and resilience within a humanitarian supply chain setting. *Int. J. Prod. Res.* **2020**, *58*, 3381–3398. [CrossRef]
3. Wagner, S.M.; Thakur-Weigold, B.; Gatti, F.; Stumpf, J. Measuring and improving the impact of humanitarian logistics consulting. *Prod. Plan. Control* **2021**, *32*, 83–103. [CrossRef]
4. Queiroz, M.M.; Fosso Wamba, S.; De Bourmont, M.; Telles, R. Blockchain adoption in operations and supply chain man-agement: Empirical evidence from an emerging economy. *Int. J. Prod. Res.* **2021**, *59*, 6087–6103. [CrossRef]
5. Dash, M.; Shadangi, P.Y.; Muduli, K.; Luhach, A.K.; Mohamed, A. Predicting the motivators of telemedicine acceptance in COVID-19 pandemic using multiple regression and ANN approach. *J. Stat. Manag. Syst.* **2021**, *24*, 319–339. [CrossRef]
6. Sahoo, K.K.; Muduli, K.K.; Luhach, A.K.; Poonia, R.C. Pandemic COVID-19: An empirical analysis of impact on Indian higher education system. *J. Stat. Manag. Syst.* **2021**, *24*, 341–355. [CrossRef]
7. Baveja, A.; Kapoor, A.; Melamed, B. Stopping Covid-19: A pandemic-management service value chain approach. *Ann. Oper. Res.* **2020**, *289*, 173–184. [CrossRef] [PubMed]
8. Bag, S.; Yadav, G.; Wood, L.C.; Dhamija, P.; Joshi, S. Industry 4.0 and the circular economy: Resource melioration in logistics. *Resour. Policy* **2020**, *68*, 101776. [CrossRef]
9. Schiffling, S.; Hannibal, C.; Fan, Y.; Tickle, M. Coopetition in temporary contexts: Examining swift trust and swift distrust in humanitarian operations. *Int. J. Oper. Prod. Manag.* **2020**, *40*, 1449–1473. [CrossRef]
10. Chou, K.P.; Prasad, M.; Lin, Y.Y.; Joshi, S.; Lin, C.-T.; Chang, J.Y. Takagi-Sugeno-Kang type collaborative fuzzy rule based system. In Proceedings of the 2014 IEEE Symposium on Computational Intelligence and Data Mining (CIDM), Orlando, FL, USA, 9–12 December 2014; pp. 315–320.
11. Zwitter, A.; Boisse-Despiaux, M. Blockchain for humanitarian action and development aid. *J. Int. Humanit. Action* **2018**, *3*, 16. [CrossRef]
12. Banomyong, R.; Varadejsatitwong, P.; Oloruntoba, R. A systematic review of humanitarian operations, humanitarian logistics and humanitarian supply chain performance literature 2005 to 2016. *Ann. Oper. Res.* **2019**, *283*, 71–86. [CrossRef]
13. Gupta, P.K.; Kumar, A.; Joshi, S. A review of knowledge, attitude, and practice towards COVID-19 with future directions and open challenges. *J. Public Aff.* **2020**, *21*, e2555. [CrossRef]
14. Joshi, S. E-Supply Chain Collaboration and Integration: Implementation Issues and Challenges. In *E-Logistics and E-Supply Chain Management: Applications for Evolving Business*; IGI Global: Hershey, PA, USA, 2013.
15. Joshi, R.; Joshi, S. Assessing the Readiness of Farmers towards Cold Chain Management: Evidences from India. In *Designing and Implementing Global Supply Chain Management*; IGI Global: Hershey, PA, USA, 2016; pp. 219–235.
16. Joshi, S. *Designing and Implementing Global Supply Chain Management*; IGI Global: Hershey, PA, USA, 2015.
17. Joshi, S.; Singh, R.K.; Sharma, M. Sustainable Agri-food Supply Chain Practices: Few Empirical Evidences from a Developing Economy. *Glob. Bus. Rev.* **2020**, 0972150920907014. [CrossRef]
18. Kovács, G.; SigalaI, F. Lessons learned from humanitarian logistics to manage supply chain disruptions. *J. Supply Chain Manag.* **2021**, *57*, 41–49. [CrossRef]
19. Kusumastuti, R.D.; Arviansyah, A.; Nurmala, A.; Wibowo, S.S. Knowledge management and natural disaster preparedness: A systematic literature review and a case study of East Lombok, Indonesia. *Int. J. Disaster Risk Reduct.* **2021**, *58*, 102223. [CrossRef]
20. Ab Malik, M.H.; Omar, E.N.; Maon, S.N. Humanitarian Logistics: A Disaster Relief Operations Framework During Pandemic Covid-19 in Achieving Healthy Communities. *Adv. Bus. Res. Int. J.* **2020**, *6*, 101–113. [CrossRef]
21. Joshi, S. Social network analysis in smart tourism-driven service distribution channels: Evidence from tourism supply chain of Uttarakhand, India. *Int. J. Digit. Cult. Electron. Tour.* **2018**, *2*, 255–272. [CrossRef]
22. Joshi, S.; Sharma, M.; Kler, R. Modeling Circular Economy Dimensions in Agri-Tourism Clusters: Sustainable Performance and Future Research Directions. *Int. J. Math. Eng. Manag. Sci.* **2020**, *5*, 1046–1061. [CrossRef]
23. Yu, L.; Zhang, C.; Jiang, J.; Yang, H.; Shang, H. Reinforcement learning approach for resource allocation in humanitarian logistics. *Expert Syst. Appl.* **2021**, *173*, 114663. [CrossRef]
24. Kunz, N.; Gold, S. Sustainable humanitarian supply chain management—Exploring new theory. *Int. J. Logist. Res. Appl.* **2017**, *20*, 85–104. [CrossRef]
25. Kelman, I. COVID-19: What is the disaster? *Soc. Anthropol.* **2020**, *28*, 296–297. [CrossRef] [PubMed]
26. Seddighi, H. COVID-19 as a Natural Disaster: Focusing on Exposure and Vulnerability for Response. *Disaster Med. Public Health Prep.* **2020**, *14*, e42–e43. [CrossRef]
27. Li, J.; An, Y.; Wang, L.; Zhang, Y. Combating the COVID-19 pandemic: The role of disaster experience. *Res. Int. Bus. Finance* **2021**, *60*, 101581. [CrossRef] [PubMed]
28. Schiffling, S.; Hannibal, C.; Tickle, M.; Fan, Y. The implications of complexity for humanitarian logistics: A complex adaptive systems perspective. *Ann. Oper. Res.* **2020**, 1–32. [CrossRef]
29. Muggy, L.; Stamm, J.L.H. Decentralized beneficiary behavior in humanitarian supply chains: Models, performance bounds, and coordination mechanisms. *Ann. Oper. Res.* **2019**, *284*, 333–365. [CrossRef]

30. Joshi, S.; Sharma, M.; Singh, R.K. Performance Evaluation of Agro-tourism Clusters using AHP–TOPSIS. *J. Oper. Strat. Plan.* **2020**, *3*, 7–30. [CrossRef]
31. Charles, A.; Lauras, M.; Van Wassenhove, L.N.; Dupont, L. Designing an efficient humanitarian supply network. *J. Oper. Manag.* **2016**, *47-48*, 58–70. [CrossRef]
32. Queiroz, M.M.; Ivanov, D.; Dolgui, A.; Wamba, S.F. Impacts of epidemic outbreaks on supply chains: Mapping a research agenda amid the COVID-19 pandemic through a structured literature review. *Ann. Oper. Res.* **2020**, *238*, 329–354. [CrossRef]
33. Thompson, D.D.; Anderson, R. The COVID-19 response: Considerations for future humanitarian supply chain and logistics management research. *J. Humanit. Logist. Supply Chain Manag.* **2021**, *11*, 157–175. [CrossRef]
34. Kamble, S.S.; Gunasekaran, A.; Parekh, H.; Joshi, S. Modeling the internet of things adoption barriers in food retail supply chains. *J. Retail. Consum. Serv.* **2019**, *48*, 154–168. [CrossRef]
35. Ivanov, D.; Dolgui, A. Viability of intertwined supply networks: Extending the supply chain resilience angles towards survivability. A position paper motivated by COVID-19 outbreak. *Int. J. Prod. Res.* **2020**, *58*, 2904–2915. [CrossRef]
36. Joshi, S.; Sharma, M. Impact of sustainable Supply Chain Management on the Performance of SMEs during COVID-19 Pandemic: An Indian Perspective. *Int. J. Logist. Econ. Glob.* **2021**. Available online: http://hdl.handle.net/10453/152609 (accessed on 22 December 2021).
37. Joshi, S.; Sharma, M. Prolonging retailer-supplier relationship: A study of retail firms during pandemic COVID-19. *Int. J. Logist. Econ. Glob.* **2021**. Available online: http://hdl.handle.net/10453/152610 (accessed on 22 December 2021).
38. Akhtar, P.; Marr, N.; Garnevska, E. Coordination in humanitarian relief chains: Chain coordinators. *J. Humanit. Logist. Supply Chain Manag.* **2012**, *2*, 85–103. [CrossRef]
39. Joshi, S.; Sharma, M. Digital technologies (DT) adoption in agri-food supply chains amidst COVID-19: An approach to-wards food security concerns in developing countries. *J. Glob. Oper. Strateg. Source.* **2021**. [CrossRef]
40. Sharma, M.; Joshi, S.; Luthra, S.; Kumar, A. Managing disruptions and risks amidst COVID-19 outbreaks: Role of blockchain technology in developing resilient food supply chains. *Oper. Manag. Res.* **2021**, 1–14. [CrossRef]
41. Shanker, S.; Barve, A.; Muduli, K.; Kumar, A.; Garza-Reyes, J.A.; Joshi, S. Enhancing resiliency of perishable product supply chains in the context of the COVID-19 outbreak. *Int. J. Logist. Res. Appl.* **2021**, 1–25. [CrossRef]
42. Malmir, B.; Zobel, C.W. An applied approach to multi-criteria humanitarian supply chain planning for pandemic response. *J. Humanit. Logistics. Supply Chain. Manag.* **2021**, *11*, 320–346. [CrossRef]
43. Joshi, S.; Sharma, M. Social capital in the Asia Pacific: Examples from the services industry. *Asia Pac. Bus. Rev.* **2018**, *25*, 457–458. [CrossRef]
44. Tripathi, G.; Joshi, S. Creating Competitive Advantage through Sustainable Supply Chains: A Theoretical Framework for the Assessment of Practices, Dynamic Capabilities, and Enterprise Performance of Manufacturing Firms. *Int. J. Recent Technol. Eng.* **2019**, *8*, 7863–7875.
45. Ivanov, D. Lean resilience: AURA (Active Usage of Resilience Assets) framework for post-COVID-19 supply chain management. *Int. J. Logist. Manag.* **2021**. [CrossRef]
46. Swain, S.; Peter, O.; Adimuthu, R.; Muduli, K. BlockChain Technology for Limiting the Impact of Pandemic: Challenges and Prospects. In *Computational Modelling and Data Analysis in COVID-19 Research*; CRC Press: Boca Raton, FL, USA, 2021; pp. 165–186.
47. Joshi, S.; Sharma, M.; Bisht, P.; Singh, S. Explaining the Factors Influencing Consumer Perception, Adoption Readiness, and Perceived Usefulness toward Digital Transactions: Online Retailing Experience of Millennials in India. *J. Oper. Strat. Plan.* **2021**, *4*, 202–223. [CrossRef]
48. Blecken, A. *Humanitarian Logistics: Modelling Supply Chain Processes of Humanitarian Organizations*; Haupt Verlag AG: Bern, Switzerland, 2010; p. 18.
49. Joshi, S.; Sharma, M.; Kumar, S.; Pant, M.K. Co-Creation Among Small Scale Tourism Firm: Role of Information Com-munication and Technology in Productivity and Sustainability. *Int. J. Strateg. Inf. Technol. Appl.* **2018**, *9*, 1–14. [CrossRef]
50. Joshi, S.; Sharma, M.; Rathi, S. Forecasting in Service Supply Chain Systems: A State-of-the-Art Review Using Latent Semantic Analysis. *Adv. Bus. Manag. Forecast.* **2017**, *12*, 181–212. [CrossRef]
51. Prasad, M.; Li, D.L.; Lin, C.T.; Prakash, S.; Singh, J.; Joshi, S. Designing Mamdani-Type Fuzzy Reasoning for Visualizing Prediction Problems Based on Collaborative Fuzzy Clustering. *IAENG Int. J. Comput. Sci.* **2015**, *42*.
52. Sharma, M.; Joshi, S. Barriers to blockchain adoption in health-care industry: An Indian perspective. *J. Glob. Oper. Strat. Sourc.* **2021**, *14*, 134–169. [CrossRef]
53. Polater, A. Dynamic capabilities in humanitarian supply chain management: A systematic literature review. *J. Humanit. Logist. Supply Chain Manag.* **2020**, *11*, 46–80. [CrossRef]
54. Ertem, M.A.; İşbilir, M.; Arslan, A.Ş. Review of intermodal freight transportation in humanitarian logistics. *Eur. Transp. Res. Rev.* **2017**, *9*, 10. [CrossRef]
55. Penna, P.H.V.; Santos, A.C.; Prins, C. Vehicle routing problems for last mile distribution after major disaster. *J. Oper. Res. Soc.* **2017**, *69*, 1254–1268. [CrossRef]
56. De Camargo, J.A.; Mendonça, P.S.M.; Oliveira, J.; Jabbour, C.J.C.; Jabbour, A.B.L.D.S. Giving voice to the silent: A framework for understanding stakeholders' participation in socially-oriented initiatives, community-based actions and humanitarian operations projects. *Ann. Oper. Res.* **2019**, *283*, 143–158. [CrossRef]

57. Mannakkara, S.; Wilkinson, S.; Potangaroa, R. *Resilient Post Disaster Recovery through Building Back Better*; CRC Press: Boca Raton, FL, USA, 2018.
58. Lopez, A.; de Perez, E.C.; Bazo, J.; Suarez, P.; van den Hurk, B.; van Aalst, M. Bridging forecast verification and human-itarian decisions: A valuation approach for setting up action-oriented early warnings. *Weather. Clim. Extrem* **2020**, *27*, 100167. [CrossRef]
59. Rana, P.; Joshi, S. Management Practices for Sustainable Supply Chain and Its Impact on Economic Performance of SMEs: An Analytical Study of Uttarakhand State, India. *Int. J. Manag.* **2020**, *11*, 346–354.
60. Janssen, M.; Van Der Voort, H. Agile and adaptive governance in crisis response: Lessons from the COVID-19 pandemic. *Int. J. Inf. Manage.* **2020**, *55*, 102180. [CrossRef] [PubMed]
61. Dash, P.; Punia, M. Governance and disaster: Analysis of land use policy with reference to Uttarakhand flood 2013, India. *Int. J. Disaster Risk Reduct.* **2019**, *36*, 101090. [CrossRef]
62. Sharma, M.; Joshi, S. Online Advertisement Using Web Analytics Software. *Int. J. Bus. Anal.* **2020**, *7*, 13–33. [CrossRef]
63. Sigala, I.F.; Kettinger, W.J.; Wakolbinger, T. Digitizing the field: Designing ERP systems for Triple-A humanitarian supply chains. *J. Humanit. Logist. Supply Chain Manag.* **2020**, *10*, 231–260. [CrossRef]
64. Gavidia, J.V. A model for enterprise resource planning in emergency humanitarian logistics. *J. Humanit. Logist. Supply Chain Manag.* **2017**, *7*, 246–265. [CrossRef]
65. Sharma, M.; Joshi, S. Digital supplier selection reinforcing supply chain quality management systems to enhance firm's performance. *TQM J.* **2020**. [CrossRef]
66. Goniewicz, K.; Khorram-Manesh, A.; Hertelendy, A.J.; Goniewicz, M.; Naylor, K.; Burkle, F.M., Jr. Current Response and Management Decisions of the European Union to the COVID-19 Outbreak: A Review. *Sustainability* **2020**, *12*, 3838. [CrossRef]
67. Sharma, M.; Joshi, S.; Govindan, K. Issues and solutions of electronic waste urban mining for circular economy transition: An Indian context. *J. Environ. Manag.* **2021**, *290*, 112373. [CrossRef]
68. Meier, P. *Digital Humanitarians: How Big Data is Changing the Face of Humanitarian Response*; CRC Press: Boca Raton, FL, USA, 2015.
69. Sharma, M.; Joshi, S.; Kumar, A. Assessing enablers of e-waste management in circular economy using DEMATEL method: An Indian perspective. *Environ. Sci. Pollut. Res.* **2020**, *27*, 13325–13338. [CrossRef]
70. Gupta, S.; Altay, N.; Luo, Z. Big data in humanitarian supply chain management: A review and further research directions. *Ann. Oper. Res.* **2019**, *283*, 1153–1173. [CrossRef]
71. Sharma, M.; Joshi, S.; Luthra, S.; Kumar, A. Analysing the Impact of Sustainable Human Resource Management Practices and Industry 4.0 Technologies Adoption on Employability Skills. *Int. J. Manpow.* **2021**. Available online: http://repository.londonmet.ac.uk/6738/ (accessed on 16 December 2021).
72. Sharma, M.; Luthra, S.; Joshi, S.; Kumar, A. Implementing challenges of artificial intelligence: Evidence from public manufacturing sector of an emerging economy. *Gov. Inf. Q.* **2021**, 101624. [CrossRef]
73. Hart, O.E.; Halden, R.U. Modeling wastewater temperature and attenuation of sewage-borne biomarkers globally. *Water Res.* **2020**, *172*, 115473. [CrossRef] [PubMed]
74. Sharma, M.; Luthra, S.; Joshi, S.; Kumar, A. Accelerating retail supply chain performance against pandemic disruption: Adopting resilient strategies to mitigate the long-term effects. *J. Enterp. Inf. Manag.* **2021**, *34*, 1844–1873. [CrossRef]
75. Ishiwatari, M.; Koike, T.; Hiroki, K.; Toda, T.; Katsube, T. Managing disasters amid COVID-19 pandemic: Approaches of response to flood disasters. *Prog. Disaster Sci.* **2020**, *6*, 100096. [CrossRef]
76. Singh, R.K.; Joshi, S.; Sharma, M. Modelling Supply Chain Flexibility in the Indian Personal Hygiene Industry: An ISM-Fuzzy MICMAC Approach. *Glob. Bus. Rev.* **2020**, 0972150920923075. [CrossRef]
77. Bhattacharya, S.; Hasija, S.; Van Wassenhove, L.N. Designing Efficient Infrastructural Investment and Asset Transfer Mechanisms in Humanitarian Supply Chains. *Prod. Oper. Manag.* **2014**, *23*, 1511–1521. [CrossRef]
78. Altay, N.; Narayanan, A. Forecasting in humanitarian operations: Literature review and research needs. *Int. J. Forecast.* **2020**. [CrossRef]
79. Biswal, J.N.; Muduli, K.; Satapathy, S. Critical Analysis of Drivers and Barriers of Sustainable Supply Chain Management in Indian Thermal Sector. *Int. J. Procure. Manag.* **2017**, *10*, 411–430. [CrossRef]
80. Altay, N.; Kovács, G.; Spens, K. The evolution of humanitarian logistics as a discipline through a crystal ball. *J. Humanit. Logist. Supply Chain Manag.* **2021**, *11*, 577–584. [CrossRef]
81. Peter, O.; Swain, S.; Muduli, K.; Ramasamy, A. *IoT in Combating Covid 19 Pandemics: Lessons for Developing Countries, Assessing COVID-19 and Other Pandemics and Epidemics using Computational Modelling and Data Analysis*; Springer: Berlin/Heidelberg, Germany, 2021; pp. 113–132.
82. Sharma, M.; Joshi, S. Brand sustainability among young consumers: An AHP-TOPSIS approach. *Young Consum.* **2019**, *20*, 314–337. [CrossRef]
83. Muduli, K.; Barve, A. Analysis of critical activities for GSCM implementation in mining supply chains in India using fuzzy analytical hierarchy process. *Int. J. Bus. Excel.* **2015**, *8*, 767. [CrossRef]
84. Sharma, M.; Joshi, S.; Kannan, D.; Govindan, K.; Singh, R.; Purohit, H.C. Internet of Things (IoT) adoption barriers of smart cities' waste management: An Indian context. *J. Clean. Prod.* **2020**, *270*, 122047. [CrossRef]
85. Seker, S.; Zavadskas, E.K. Application of Fuzzy DEMATEL Method for Analyzing Occupational Risks on Construction Sites. *Sustainability* **2017**, *9*, 2083. [CrossRef]

86. Prakash, S.; Joshi, S.; Bhatia, T.; Sharma, S.; Samadhiya, D.; Shah, R.R.; Kaiwartya, O.; Prasad, M. Characteristic of enterprise collaboration system and its implementation issues in business management. *Int. J. Bus. Intell. Data Min.* **2020**, *16*, 49. [CrossRef]
87. Kannan, G.; Muduli, K.; Devika, K.; Barve, A. Investigation of influential strength of factors on GSCM adoption in mining in-dustries operating in India. *Resour. Conserv. Recycl.* **2016**, *107*, 185–194.
88. Brem, A.; Viardot, E.; Nylund, P.A. Implications of the coronavirus (COVID-19) outbreak for innovation: Which technologies will improve our lives? *Technol. Forecast. Soc. Change* **2021**, *163*, 120451. [CrossRef]
89. Fernandez-Luque, L.; Imran, M. Humanitarian health computing using artificial intelligence and social media: A narrative literature review. *Int. J. Med. Inform.* **2018**, *114*, 136–142. [CrossRef]
90. Quarshie, A.M.; Leuschner, R. Interorganizational Interaction in Disaster Response Networks: A Government Perspective. *J. Supply Chain Manag.* **2020**, *56*, 3–25. [CrossRef]
91. Centobelli, P.; Cerchione, R.; Del Vecchio, P.; Oropallo, E.; Secundo, G. Blockchain technology for bridging trust, traceability and transparency in circular supply chain. *Inf. Manag.* **2021**, 103508. [CrossRef]
92. Centobelli, P.; Cerchione, R.; Del Vecchio, P.; Oropallo, E.; Secundo, G. Blockchain technology design in accounting: Game changer to tackle fraud or technological fairy tale? *Account. Audit. Account. J.* **2021**. [CrossRef]

sustainability

MDPI

Article

Blockchain Technology for Enhancing Traceability and Efficiency in Automobile Supply Chain—A Case Study

Nesrin Ada [1], Manavalan Ethirajan [2], Anil Kumar [3], Vimal K.E.K [4], Simon Peter Nadeem [5], Yigit Kazancoglu [6] and Jayakrishna Kandasamy [2,*]

1 Department of Business Administration, Ege University, Izmir 35100, Turkey; nesrin.ada@ege.edu.tr
2 School of Mechanical Engineering, VIT University, Vellore 632014, Tamil Nadu, India; mailtomanav@gmail.com
3 Guildhall School of Business and Law, London Metropolitan University, London EC2M 6SQ, UK; A.Kumar@londonmet.ac.uk
4 Department of Mechanical Engineering, National Institute of Technology, Patna 800005, Bihar, India; vimalkek.engr@gmail.com
5 Centre for Supply Chain Improvement, University of Derby, Derby SK17 6RY, UK; s.nadeem@derby.ac.uk
6 Department of Logistics Management, Yasar University, Izmir 35100, Turkey; yigit.kazancoglu@yasar.edu.tr
* Correspondence: jayakrishna.k@vit.ac.in; Tel.: +91-98-94968596

Abstract: A robust traceability system would help organizations in inventory optimization reduce lead time and improve customer service and quality which further enables the organizations to be a leader in their industry sector. This research study analyzes the challenges faced by the automotive industry in its supply chain operations. Further, the traceability issues and waiting time at different nodes of the supply chain are considered to be priority issues that affect the overall supply chain efficiency in the automotive supply chain. After studying the existing blockchain architectures and their implementation methodology, this study proposes a new blockchain-based architecture to improve traceability and reduce waiting time for the automotive supply chain. A hyper ledger fabric-based blockchain architecture is developed to track the ownership transfers in inbound and outbound logistics. The simulation results of the proposed hyper ledger fabric-based blockchain architecture show that there is an improvement in the traceability of items at different nodes of the supply chain that enhances the Inventory Quality Ratio (IQR) and the mean waiting time is reduced at the factory, wholesaler, and retailer, which thereby improves the overall supply chain efficiency. The blockchain embedded supply chain is more capable to eliminate the risks and uncertainties associated with the automotive supply chain. The benefits of adopting blockchain technology in the automotive supply chain are also described. The developed blockchain-based framework is capable to get more visibility into goods movement and inventory status in automotive supply chains.

Keywords: automotive supply chains; blockchain; simulation; case study; Industry 4.0

Citation: Ada, N.; Ethirajan, M.; Kumar, A.; K.E.K, V.; Nadeem, S.P.; Kazancoglu, Y.; Kandasamy, J. Blockchain Technology for Enhancing Traceability and Efficiency in Automobile Supply Chain—A Case Study. *Sustainability* **2021**, *13*, 13667. https://doi.org/10.3390/su132413667

Academic Editors: Helena Carvalho and Ripon Kumar Chakrabortty

Received: 18 October 2021
Accepted: 7 December 2021
Published: 10 December 2021

Publisher's Note: MDPI stays neutral with regard to jurisdictional claims in published maps and institutional affiliations.

Copyright: © 2021 by the authors. Licensee MDPI, Basel, Switzerland. This article is an open access article distributed under the terms and conditions of the Creative Commons Attribution (CC BY) license (https://creativecommons.org/licenses/by/4.0/).

1. Introduction

As industries around the globe expand, their supply chains have become complex and defragmented [1–3]. Despite enormous investments being made to improve the part tracking or value tracking in supply chains, most companies still have a limited amount of part tracking mechanisms [4–6]. Many organizations have a significant gap between systems employed within a company and across companies. A traceability system is necessary to obtain a reduction in costs, waiting time, and an overall improvement in quality and customer service, which would further enable organizations to develop a competitive advantage [7,8]. Moreover, consumers nowadays are also interested in knowing whether the product they received comes from an ethical background. In recent days the importance of automation within the supply chain process has increased [9]. To automate the supply chain process, multiple systems need to be integrated which essentially means the volume of data produced. This data is essential to track the products and their status at each point to ensure quality [10].

The automotive industry is an important sector that also drives the economy of a nation [11]. Today, organizations are facing different challenges in this sector. Similar to many sectors, the automotive industry also increased its global presence which resulted in frequent part movements across the globe. Firstly, manufacturers are finding difficulty in tracing the in-transit parts, in-house production, and out for delivery products [12,13]. It remains an important challenge for the entire supply chain. Secondly, overcapacity is another problem that results in overspending that affects the overall efficiency of supply chain operations [14]. Customers are concerned about the sustainability of the parts and interested in knowing the origin of the parts. In the automotive industry, it becomes imperative to trace the components and semi-finished goods used in a particular vehicle or product family [15,16]. All the stakeholders of the supply chain lack a common information access framework, making this process difficult to execute in real-time as well as for them to exchange information in real-time [17,18]. The supply chain organizations also find it challenging to maintain the right Inventory Quality Ratio (IQR) across different nodes of the supply chain and that leads to traceability issues. Further, there is unplanned production downtime due to stock out of raw materials and machine breakdowns. The unplanned downtime affects the lead time and the waiting time for the customer is increased. Thus, impacts the supply chain efficiency [19,20].

One of the technologies that are emerging with Industry 4.0 is blockchain. Blockchain has the potential to address some of the issues faced by the automotive industry. Blockchain is a better solution for traceability issues as it can share information across supply chain networks with increased security of information. Each supply chain member can see the same information on a product's lifecycle [21]. Blockchain can potentially affect supply chain parameters such as waiting time, cost, risk reduction, speed, quality, dependability, flexibility, etc. [22,23]. With today's complex supply chain networks, the interactions and transactions among these stakeholders should exist on an immutable ledger/database system that is shared, secured, and can provide permission accessibility. A shared blockchain-based system facilitates increased transparency which enables seamless transactions and improved visibility [24]. One of the benefits of blockchain in the automotive industry is traceability that includes part provenance, vehicle tracking, improved inbound plant logistics, etc. [25]. However, blockchain adoption in the supply chain is still at its nascent stage.

The blockchain is a decentralized database with a collaborative network that functions as a ledger for maintaining secured transactional data [26]. While the applicability of blockchain technology showcases to have a considerably strong case for changing many aspects of the working of the automotive industry functioning, the automotive sector has just begun to scratch the surface of blockchain applications about its operations. Figure 1 shows a simple automotive supply chain depicting how product flow happens in the supply chain from suppliers to customers. Considering the volume of components involved in the automotive supply chain, blockchain can help manage its operations. Companies are still exploring ways to enhance the working of their supply chains as well as embrace the change that blockchain technology has to offer.

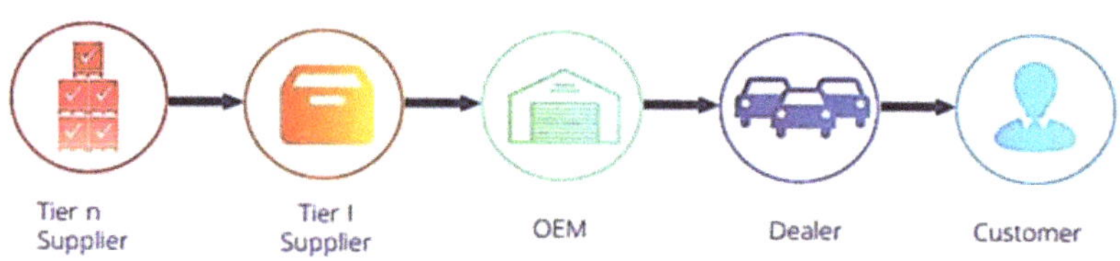

Figure 1. A typical automotive supply chain.

In the current supply chain, the trust among the supply chain partners is said to be less even though information systems are deployed. Organizations are focusing on a good relationship with other supply chain members to improve their trust however there is no guarantee to validate the information is accurate as some supply chain members may share false information with or without intention which impacts the entire eco-system. This article tries to implement blockchain technology in an automotive contemporary organization to improve supply chain operations. The literature shows that blockchain in the automotive supply chain is still in the initial stages and there is scope to study this technology and explore the opportunities to implement it in the automotive sector.

Based on the literature, the automotive supply chain with blockchain is not completely explored to overcome traceability issues and improve efficiency. Besides, the automotive supply chain requires a trustable eco-system as it involves multiple parties and stakeholders. Thus, the study on the automotive supply chain is essential to add more value to manufacturers and explore opportunities with blockchain technology. Therefore, this study intends to contribute to the blockchain literature for the automotive supply chain by addressing the following research questions.

RQ1. How an automotive supply chain traceability system based on blockchain technology can be established?
RQ2. What are the benefits of adopting a blockchain-based system in the automotive supply chain?

To address the mentioned research questions, the following objectives are defined:

- To identify different traceability issues at various nodes of an automotive supply chain.
- To develop a new blockchain architecture for the automotive supply chain and improve supply chain traceability issues and reduce waiting time, thereby improving supply chain operational efficiency.
- To explain the implications of implementing blockchain across the automotive supply chain.

By addressing the research questions and objectives, this article contributes to the blockchain literature by identifying the issues in the automotive supply chain. The result can help organizations to understand the impact of blockchain particularly for automotive manufacturers who work in a complicated supply chain network with a high volume of parts involved.

The article is constructed as follows. Section 2 provides the literature on the issues faced in the traditional supply chain, how blockchain applications help in the supply chain. Further, it discusses the research gap based on the literature study. Section 3 explains the research methodology. Section 4 describes the case organization details, application of the proposed framework, and presents the simulation results of the blockchain framework. Section 5 discusses the results obtained and explains the theoretical and practical implications. Finally, the key outcomes and conclusions are summarized in Section 6.

2. Literature Review

The existing literature was reviewed to identify different methods used for supply chain traceability and their limitations. The investigation was carried out using different keywords related to supply chain, supply chain, data management systems, traceability, synchronization, blockchain architecture, framework model, etc. in various databases such as a web of science, IEEE Xplore, Springer link, Scopus, Taylor and Francis, Science Direct, Wiley, Emerald. Concerning the reported domain of research, 60 research papers, four reports, and three white papers were shortlisted considering the aim and scope of the study from an enormous amount of literature available relevant to the above-stated keywords.

2.1. Issues in Traditional Supply Chain

Various supply chain issues related to automotive supply were reviewed. Bonilla et al., (2018) analyzed traditional supply chain issues and analyzed Industry 4.0 technologies to improve sustainability [27]. Di Vaio and Varriale (2020) reviewed the issues in the supply chain related to the airport industry and explored blockchain to improve sustainable performance [28]. Junaid et al., (2020) assessed supply chain risk using AHP and TOPSIS framework for the automotive industry [29]. González-Benito et al., (2013) conducted a bibliometric analysis of 404 publications which is focused on the automotive industry [30]. Maro et al., (2018) provided a comprehensive overview of the challenges involved in effective traceability as well as solutions in the automotive domain while contrasting them with those found in general literature [31]. Papetti et al., (2019) aimed at developing a platform for concurrently supporting supply chain traceability as well as eco-sustainability in the leather shoe supply chain [32]. Ferriols et al., (2013) proposed eight steps for developing a hierarchy in a supply chain company [33].

When multiple parties are involved, few manual processes happen across the supply chain which leads to incorrect data reconciliation and a drop in efficiency [34,35]. In addition, there are chances for mixing counterfeit products when there is no direct visibility for Original Equipment Manufacturers (OEM) with tier-2 suppliers [36,37]. The problem of differential pricing comes into the picture as they generally prefer concealing their pricing since this allows them to pay lower prices when outsourcing to developing countries [38–40]. As numerous parties are involved, mediating between these parties can pose a critical problem for logistics providers such as slowing down the delivery of services and creating a large overhead for logistics. Furthermore, a centralized mediator of these parties can misuse power to prioritize some parties over others. As quality and compliance issues occur, procuring a replacement for defective parts is long-drawn and unfeasible in some cases. Lean "on-demand" manufacturing falls flat in a situation where natural disasters and socio-economic problems are common. For example, Japan (frequently affected by earthquakes) has outsourced most of its supply chain logistics to other countries. Hence, it is safe to say that in some cases, companies rely on their disruptive network to function effectively. But as a central mediator for parties is required which centralizes power in the hands of a few may also be a gateway to misuse the resources. As the number of interacting parties increases, there is a proportional increase in middlemen. It may lead to fraud and slow down the supply without contributing to the network. In a centralized network validating identity vendors and checking for tampering by middlemen is cumbersome and, in some cases, outright impossible. Table 1 summarizes supply chain issues and descriptions.

Table 1. Typical supply chain issues and description.

Issue	Description	References
Tracking the history of any product	Validating identity vendors and checking for tampering by middlemen is not possible.	[2,41,42]
Differential Pricing	Companies prefer keeping their pricing a secret since this allows them to pay lower prices when outsourcing to developing countries	[43,44]
Numerous Parties Involved	Mediating between so many parties can be a big problem for logistics providers, slowing down the delivery of services and creating a large overhead for logistics. Furthermore, a centralized mediator of these parties can misuse power to prefer some parties over others.	[44–47]
Quality and Compliance Issues	Procuring a replacement for defective parts is a long-drawn and uncomfortable process.	[48–51]
Inevitable disruptions	LEAN "on-demand" manufacturing falls flat in a situation where natural disasters and socio-economic problems are common. For example, Japan (frequently affected by earthquakes) has outsourced most of its supply chain logistics to other countries.	[52–54]
Centralization	A central mediator for parties is required, which centralizes power in the hands of a few and is a gateway to misuse of resources	[55–57]
Fraud by Middlemen	As the number of interacting parties increases, there is a proportional increase in middlemen. They lead to fraud and slow down the supply without adding anything to the network	[58–60]

Özceylan et al., (2016) reported on the development of a linear programming model for handling reverse material flows and integrating them with forwarding supply chains [61]. Sher et al., (2018) developed a framework to synchronize activities in a supply chain for perishable products that require a careful storage system [62]. Al-aomar and Hussain

(2018) explored issues concerning lean and sustainability issues in service-based supply chains [63]. Kalverkamp and Young (2019) developed a framework for closed-loop supply chains. Case studies of Automotive remanufacturing are considered here [64]. Masoud and Mason (2015) developed a hybrid model using a hybrid simulated annealing algorithm (HSAA) which solves the NP-hard problem. A tier-1 automotive industry supply chain was considered for the same [65].

Borgstedt et al., (2017) studied alternative powertrain systems and justified the technology shift required in the automotive industry [66]. Mathivathanan et al., (2017) developed a framework for Sustainable Supply Chain Management (SSCM) practices within the automotive industry [67]. Ambe and Badenhorst-Weiz (2011) proposed a supply chain plan that provides standard rules for implementing in a demand-driven supply chain environment [68]. Sundarakani et al., (2012) conducted a worldwide economic analysis of the automotive industry to identify the potential factors that favor and also provided recommendations for the automotive industry based in Singapore [69]. Bhattacharya et al. (2014) examined subjective parameters such as visibility, innovation in the supply chain, and collaboration among supply networks [70]. Teucke et al., (2018) studied a system that affects the production planning stages in the supply chain [71]. Queiroz and Fosso Wamba (2019) have developed a framework built on the theory of acceptance [41].

2.2. Blockchain Technology in Supply Chain

The literature on blockchain application with the supply chain was analyzed. Li et al., (2018) developed a framework that employs blockchain for systems in injection mold redesign to improve qualitative aspects such as trust, brand value, etc [42]. A case study for developing and implementing blockchain architecture in a cloud manufacturing environment and it is found that the total time required for the transaction process is significantly reduced [43]. Muzammal et al., (2019) present an application platform with the capabilities of and fast processing of data [44]. Banerjee (2018) described the benefits of integrating ERP with blockchain to improve supply chain effectiveness [45]. Atlam and Wills (2018) provided information about blockchain and IoT [46]. Min (2019) analyzed various areas of risk management and security for employing Blockchain technology [47]. Xu et al., (2019) conducted a study on blockchain and how it can be implemented in various product traceability scenarios [48]. Casado-vara et al., (2018) developed a blockchain model is for integrating it with an existing supply chain in the agriculture sector by employing smart contracts [49]. Costa et al., (2017) discuss the visibility of internal logistical processes using Radio Frequency Identification (RFID) technology. Additionally, this study identifies and discusses the main obstacles that are associated with the implementation of its proposed solution. When compared to the data management frameworks of conventional traceability tools such as RFID tagging and E-Kanban, blockchain has the potential to overcome its limitations and significantly improve supply chain operations. The data management system considering RFID tagging needs to be updated frequently as surplus amounts of variable data are produced. The E-Kanban system is slower to adapt to this changing set of data [50]. Furthermore, this system does not work ahead of schedule. While these traceability systems need their infrastructure to be set up, the data management framework of blockchain can be integrated with conventional traceability tools.

With the advent of disruptive technologies, conventional data management systems could be replaced or integrated into modular and scalable systems. The blockchain, being immutable, distributed, and decentralized in nature with its scalability and modularity, could prove an effective alternative for the replacement of data management systems of conventional traceability tools. Furthermore, it could also be integrated with them, if required. Table 2 shows the summary of data management methods used in various traceability tools.

Table 2. Summary of data management methods used in various traceability tools.

	RFID	E—Kanban	Blockchain
Features	Provides relatively accurate location over a certain range	Allows visual management	Distributed, decentralized, A system with immutable, tamper-proof nature
	Can be integrated with Enterprise resource planning (ERP) systems	Improves visibility in plant and paperless operation	Enables real-time information and enhances visibility through the entire supply chain
	Improves accuracy in logistical operations	Supports lean and helps to reduce waste	Can be integrated with RFID and E-Kanban
Limitations	Limitations in operating conditions	Need for financial investment	Requires adaption to new tech
	Less accurate/faulty detection	Centralized system	Need for financial investment
	Security and privacy issues	Can easily be tampered/modified	A seamless digital-physical link
	Can be tampered/modified	Limited scalability	Needs acceptance by every supply chain partner

Different techniques have evolved which facilitate the live tracking of parts and goods in a supply chain. However, data management frameworks of current traceability systems such as RFID and E-Kanban are centralized and lack the distributed and scalable architecture that can be used for real-time secure data sharing with all supply chain partners. While the current research on various applications of blockchain technology is still evolving, there is a limited amount of research being carried out on blockchain implementation in an automotive supply chain. Although a variety of relevant blockchain applications are being proposed, the pertinent discussions are mostly conceptual expositions, with very little or no empirical evidence of how to employ this technology at the organizational level. Blockchain-enabled supply chain, if successfully implemented, would have potential advantages because of its tamper-proof nature, the distributed structure which would further enable improvement in visibility on the inventory of the organization. Hence there is a need to develop a framework to embed blockchain in the automotive supply chain.

2.3. Research Gap

Based on the literature review, it is observed that some of the research work has been carried out on traditional supply chain issues and the application of blockchain technology. The use of blockchain for traceability issues and reducing waiting time are not reviewed for the automotive supply chain. There is a need for organizations to focus on these issues to improve overall operational efficiency. Kamble et al., (2021) reviewed blockchain technology's impact on the automotive supply chain and summarized the relationship between blockchain and sustainable supply performance [51]. Khanfar et al., (2021) reviewed the application of blockchain technology in manufacturing and suggested opportunities for new directions of research [52]. Rejeb et al., (2021) reviewed blockchain technologies in logistics and supply chain management and summarizes the existing gaps and potential research directions for future research in the blockchain [53]. Further, it appears that there are not many studies earlier that performed simulation-based analysis for traceability issues and waiting for time problems with the focus on improving operational efficiency in the automotive supply chain. Thus, studies unveil that the research on blockchain for the automotive supply chain is slightly limited and suggest exploring a more practice-oriented approach. The mentioned factors are considered as the literature gap for the present study.

3. Methodology

To formulate the methodology, different use cases concerning various traceability breaches in food and automotive manufacturing industries were reviewed and common problems faced by stakeholders especially in areas on the tracking of the component/parts were identified. Furthermore, various solution methodologies like the use of IoT sensors were reviewed. However, to govern the system and ensure the reliability of data from these sensors and different methodologies, a virtually unbreachable model is required. Hence, blockchain as a potential solution for the aforementioned problem is proposed in this study. The blockchain technology applied on various applications is reviewed to develop a solution approach by simulating a supply chain. Weak nodes or links that possess potential threats to traceability are identified within the supply chain. A blockchain model is developed to understand the risks posed by these critical nodes, thus, enabling effective integration of technology in the automotive organization. Figure 2 shows the flowchart of steps involved in the methodology.

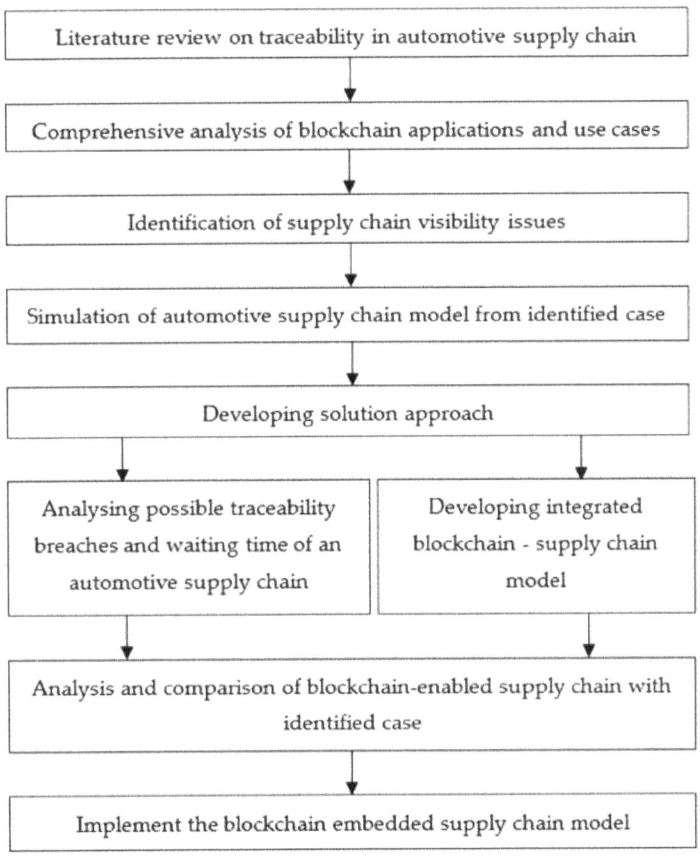

Figure 2. Research methodology framework.

4. Application of Proposed Framework in the Case Organization

4.1. Case Details

The case organization is selected based on its supply chain maturity level and present IT systems. The case organization is a leading automotive manufacturer in South India and leading manufacturers of steering and suspense systems, valve train components, friction material products such as brake linings, disc pads, clutch facings. This organization utilizes modern technologies to provide the clients a perfect fit in a variety of industry segments such as passenger vehicles, commercial vehicles, and farm tractors. The organization has put into practice world-class approaches such as ISO 9001:2015 for quality management systems and also won the prestigious 'Deming Grand Prize' for their supply chain quality practices. The contemporary organization is in the digital transformation phase to get real-time visibility on their supply chain to make quicker decisions. The organization is interested and willing to explore blockchain technology for improving supply chain operational efficiency. The organization thus selected is best suited for the study as they are positive about adopting new technologies and distinguish themselves as a global leader in the automotive sector.

4.2. Current Supply Chain Operational Challenges for Case Organization

The case organization faces challenges in maintaining the correct Inventory Quality Ratio (IQR). IQR is an approach for managing inventory to find the fast-moving items, and review the slow-moving and non-moving items. The focus is to track the active inventory and reduce the slow-moving and non-moving inventory at different nodes of the supply chain that improves the IQR. Additionally, they are times the production is stopped due to the non-availability of materials and tool breakdowns. Hence the organization is keen on improving IQR and also reducing the waiting time so that the overall operation efficiency is improved.

Based on the traceability and efficiency issues reported, the data is collected on Key Performance Indicators (KPI) for calculating IQR and Waiting time. IQR is calculated by measuring active inventory divided by the total inventory. The total inventory comprises active, slow-moving, excess, obsolete inventories. In addition, the waiting time is calculated based on the number of stops per shift and the average time lost per shift. Based on the computation, Table 3 is formulated. The calculated KPIs are validated with the stakeholders who are expertise in the area of supply chain operations belonging to production, logistics, and inventory departments.

Table 3. Evaluation on factors affecting IQR and Waiting time for the automotive supply chain.

KPI	Factors	Last 12 Months (until June 2021)											
		1	**2**	**3**	**4**	**5**	**6**	**7**	**8**	**9**	**10**	**11**	**12**
IQR (Active Inventory/Total Inventory)	Active, Slow, Excess, Obsolete Inventory (%)	62	67	72	68	65	69	74	70	63	64	71	68
Waiting time	Number of stops/shift	4	5	6	2	3	8	6	7	6	8	5	5
	Average Time lost/shift (Minutes)	10.3	12.6	15.1	5.6	7.5	20.4	13.5	16.7	14.56	20.5	12.7	13.8

Based on the analysis, a model is proposed as illustrated in Figure 3 nonetheless similar to the general automotive supply chain as depicted by Figure 1, and can be considered as an abridged version of the same. Here the first three blocks of Figure 1 are replaced by a factory, wholesaler acts as a dealer, and retailer as a customer. On arrival, the customer demands a product from a retailer and purchases the available products. In case the demand exceeds the supply, the excess is backlogged. The respective inventory levels of the retailer and wholesaler are reviewed every day. Upon reviewing the order from the customer, every member in the supply chain determines the overall demand which needs to be fulfilled like how many items are to be ordered. Different types of costs, such as carrying cost, shortage cost, and ordering cost, are calculated. Information for the simulation is given according to the case study identified, wherein the demand is randomized. The output shows the mean daily cost for all parties, as well as the distribution of waiting times for customers. The variation plots for waiting time and mean cost with max and min stock level for each party are also obtained to understand the magnitude of the variation. Anylogic software v8.4 was used to simulate the case study.

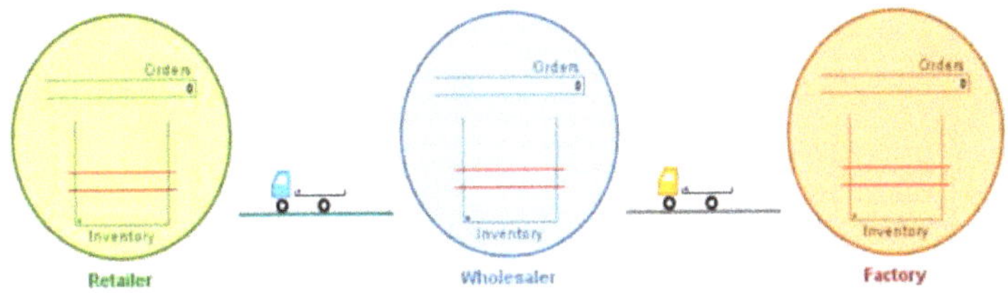

Figure 3. Auto part supply chain model.

4.3. *Application of Solution Framework*

The proposed framework is multi-layered, in that it connects various technologies as well as technical components [54]. As illustrated in Figure 4, the blockchain architecture of the distributed ledger consists of various sub-networks, each associated with specific orders and partners. In this case, the transactions are termed as per their usage in Hyperledger composer which is discussed further. It has four stakeholders namely person/customer, manufacturer, supplier, and regulator who are involved in inventory management, two assets such as order/part and vehicle that impacts the logistics, and two different transactions such as Place Order and Update Order Status that are associated with two events Place Order Event and Update Order Status Event are considered. The stakeholders in the sub-network are the parties that engage in transactions. This can also be expanded to employ an interoperable network of networks at a global scale that would be open to every supply chain partner and third-party logistics (3PL) provider. Thus, it can serve to audit the events associated with each transaction, monetary or non-monetary. There are four participants involved in the solution architecture: an index server, peers (stakeholders of the supply chain), administrative nodes, and external monitors. Considering the contemporary case organization supply chain complications, it is recommended to use a flexible blockchain framework. Thus, IBM Hyperledger Fabric is proposed for the automotive organization.

4.3.1. Blockchain Implementation Using IBM Hyperledger

Hyperledger Fabric is an open-source blockchain infrastructure, which offers a modular architecture to develop distributed systems, with an emphasis on the improvement in the reliability and performance of these distributed systems. It provides a solution approach facilitating performance at scale and simultaneously ensuring required privacy for the provision of an interoperable network of networks. The network has four participant types, two asset types, and three transaction types. The model file that defines assets, participants was developed using Hyperledger Composure Modelling Language (HCML). The script file that defines the logic of transaction executions was developed using JavaScript. A permission access control file was created to decide the access level for each stakeholder.

The index server manages the details of the participant nodes involved in the blockchain. Every participant is given a unique ID as well. The second partner comprises sub-partners or trading partners such as suppliers, manufacturers, and customers. The application node acts as a transmitter node. Essentially, it means that the application node receives the communication and transmits it to ERP. It helps to maintain the proper exchanges of information happening and minimize communication errors. Third-party systems validate the respective status of the orders and communicate them to the public ledger. This third party may also act as a regulator monitoring the transactions and handling queries. All the peer nodes deliver information to the ledger, in the form of a timestamped record of transaction events. The architecture maintains different types of events to record the status of an order as discussed below.

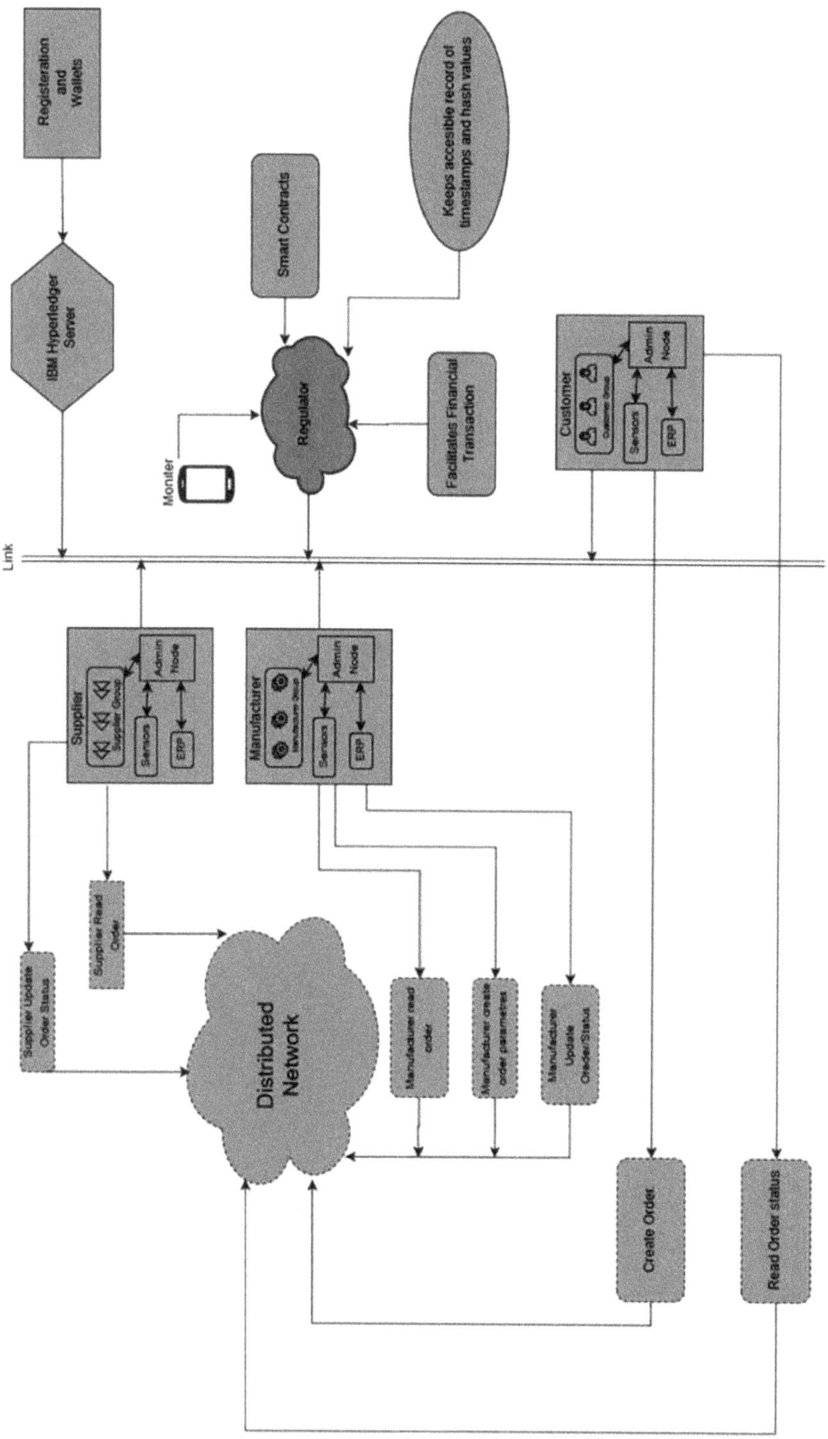

Figure 4. Architecture for blockchain integration.

4.3.2. Inception Event

The inception event shows the initiation of an order which is similar to a bitcoin transaction. This event is triggered by the customer administrative node and communicated to other members. The supply chain partner stores the information and communicates it internally. This document, however, can be tampered with in case of a security breach but the order placed in the network cannot. The information in the inception event is available only to supply chain partners. A hash value generated from this event can be verified by the customer. This specific hash value is visible to all partners.

4.3.3. Custody Event

This event is used to maintain the status of the order. It shows the current status of the order and who is responsible for it. It also shows the communication events happening between the participants. This event maintains the information in a private ledger and shares the information among supply chain members. The generated hash value is visible to all partners.

4.3.4. Monitoring Event

The monitoring event also denotes the location of the order. This event tracks the order and provides the information shared between the manufacturer and logistics provider. The generated hash value by monitoring the event is visible to all partners through a public ledger.

4.4. Simulation Results of Blockchain Framework

Figure 5 shows inventory variation (on the Y-axis) with the number of days (on the X-axis). Here a random input is given to the model which simulates for one year. As seen from Figure 5, even though the inventory at the factory is changed there is a lag before which the inventory at wholesaler and retailer is changed. This can be attributed to uncertainties occurring during information sharing between respective partners.

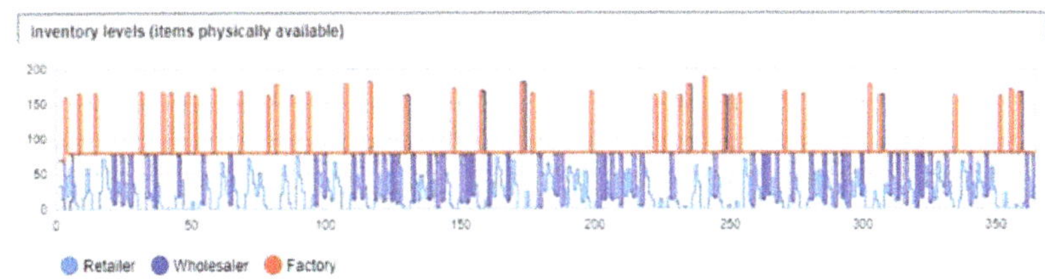

Figure 5. Inventory level before blockchain integration.

Projected Effect of Blockchain Implementation on Case Identified

Integration of blockchain would improve daily supply chain operations. Secured real-time data sharing would help to increase the level of consensus among supply chain partners, resulting in frequent order and inventory updates. This would help to lower daily costs for all the stakeholders viz. factory, wholesaler and dealer as well as lower their respective mean waiting time. Additionally, the waiting time for end customers would decrease due to the fast and responsive nature of the blockchain-enabled supply chain. The optimized simulation for a period of 1 year was carried out that produces a plot of inventory levels against the number of days. Figure 6 shows the simulation results after the integration of blockchain. It can be seen from the graphical results that though the inventory is changing randomly, the inventory of the wholesaler and retailer changes almost simultaneously as that of the factory. This can be attributed to the improved information sharing achieved by the integrated model of blockchain into the supply chain.

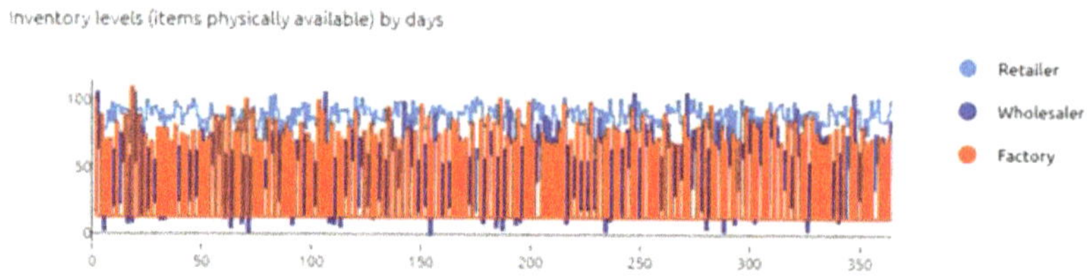

Figure 6. Inventory level after blockchain integration.

After comparing Figures 5 and 6, it is evident that there is less lag in inventory update for all the supply chain partners after blockchain implementation. This shows signs of improved communication and traceability.

Figure 7a shows more simulation results depicting the costs associated with inventory. Here, ordering, holding, and storage costs are associated with the retailer, wholesaler, and factory; the mean waiting time and the total daily cost of the supply chain are shown. Figure 7b shows the simulation result after the integration of blockchain.

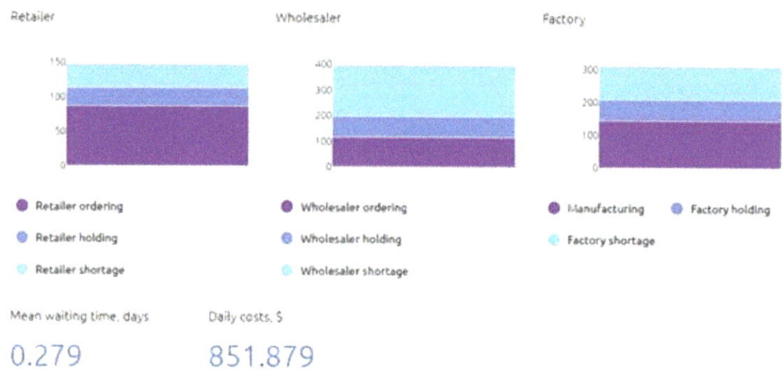

(a)

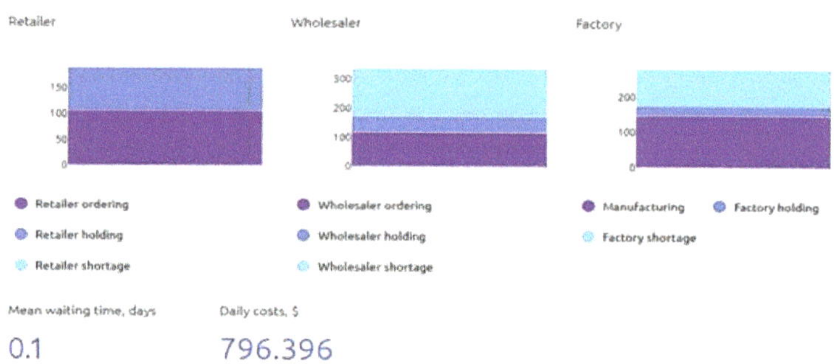

(b)

Figure 7. (**a**) Overview of daily operational costs means waiting time and total daily costs before blockchain implementation; (**b**) Overview of daily operational costs mean waiting time and total daily costs after blockchain implementation.

It can be inferred from Figure 7a,b that storage cost for the retailer is drastically decreased, however holding costs have increased. Moreover, holding costs for wholesaler and factory have decreased significantly. The mean waiting time has decreased from 0.279

to 0.1 days (i.e., 61.29% reduction in mean waiting time). Moreover, daily costs have decreased from USD 851.87 to USD 796.39 (i.e., a 6% reduction in daily costs).

Furthermore, to show the variation of factors such as daily costs and mean waiting time against the minimum and maximum costs levels, three-dimensional plots were plotted before and after the integration of blockchain. Figures 8a, 9a, 10a, 11a, 12a and 13a represent the simulation results of daily costs and mean waiting time before integrating blockchain at factory, wholesaler, and retailer. Figures 8b, 9b, 10b, 11b, 12b and 13b represent the simulation results of daily costs and mean waiting time after the integration of blockchain at factory, wholesaler, and retailer.

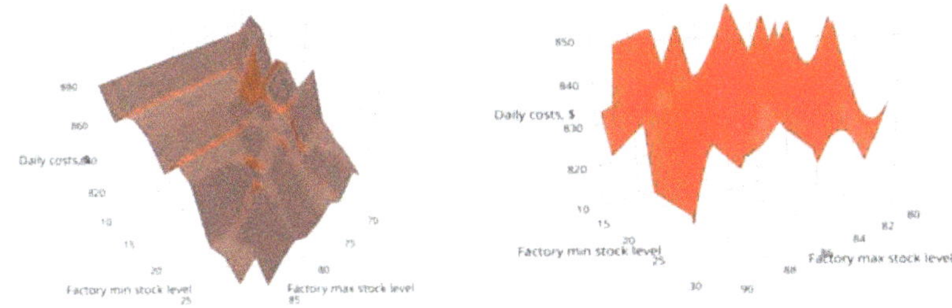

(a) (b)

Figure 8. (**a**) Factory daily costs. (**b**) Factory daily costs.

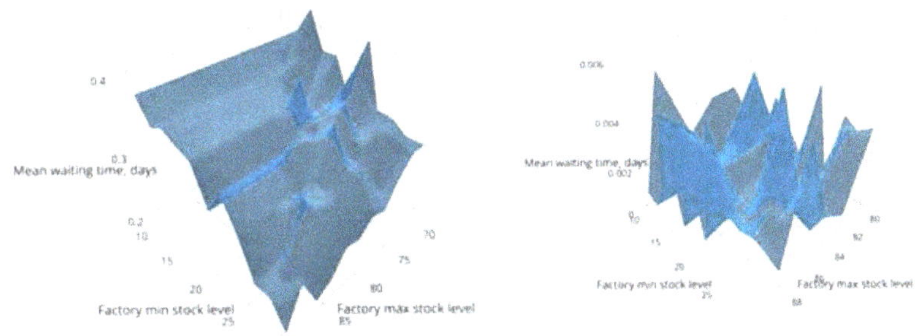

(a) (b)

Figure 9. (**a**) Factory mean waiting time. (**b**) Factory mean waiting time.

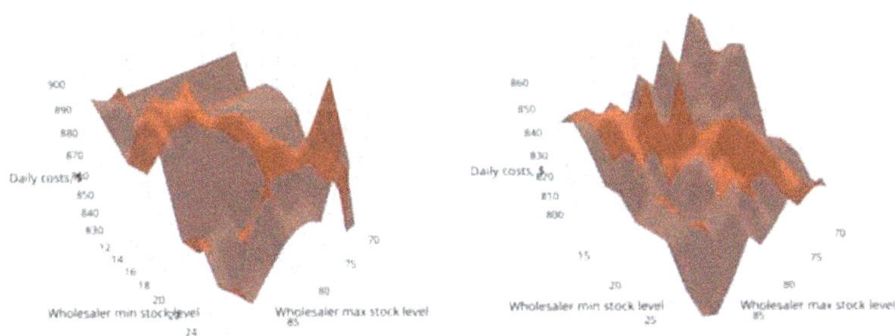

(a) (b)

Figure 10. (**a**) Wholesaler daily costs. (**b**) Wholesaler daily costs.

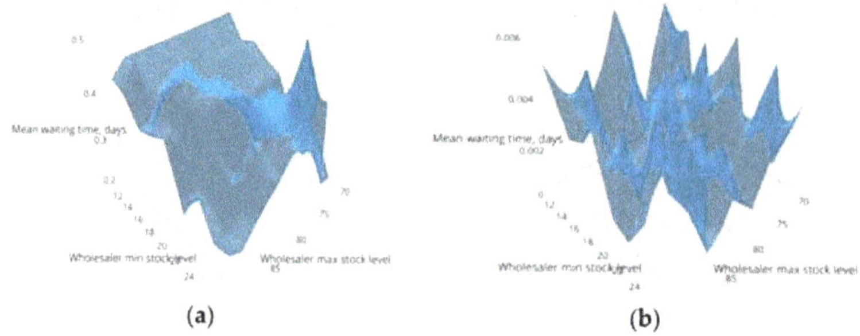

Figure 11. (**a**) Wholesaler mean waiting time. (**b**) Wholesaler mean waiting time.

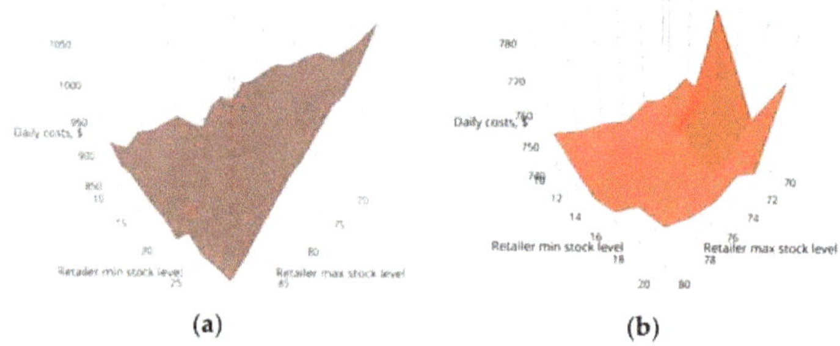

Figure 12. (**a**) Retailer daily costs. (**b**) Retailer daily costs.

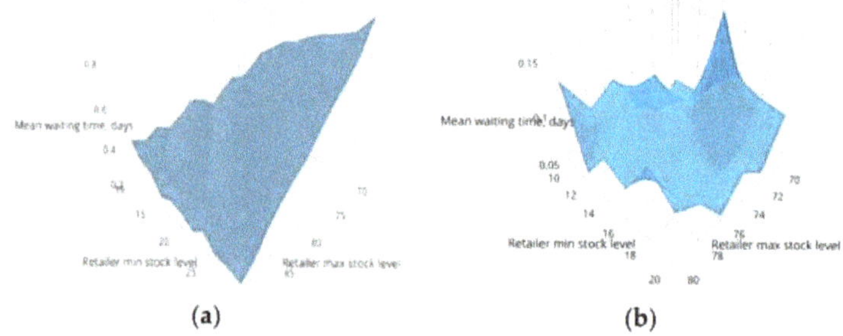

Figure 13. (**a**) Retailer mean waiting time. (**b**) Retailer mean waiting time.

Figure 8a,b depict plots that show a variation of factory daily costs before and after the integration of blockchain, respectively. Figure 9a,b illustrate the variation of mean waiting times before and after the integration of blockchain with minimum factory stock level and maximum factory stock level, respectively, which helps to measure IQR.

Figure 10a,b depict plots that show a variation of wholesaler daily costs before and after the integration of blockchain, respectively. Figure 11a,b illustrate the variation of mean waiting times before and after the integration of blockchain respectively with minimum wholesaler stock level and maximum wholesaler stock level that helps to measure IQR.

Figure 12a,b depict plots that show a variation of retailer daily costs before and after the integration of blockchain, respectively. Figure 13a,b illustrate the variation of mean waiting times before and after the integration of blockchain with minimum retailer stock level and maximum retailer stock level, respectively, which helps to measure IQR.

5. Discussion

After comparing Figures 8–13, it can be inferred that the maximum limit of mean daily costs and mean waiting time for factory, wholesaler, and retailer after employing a blockchain has significantly decreased with minimum retailer stock level and maximum retailer stock level that helps to measure IQR. The deployment of the blockchain model improves the data and information sharing between different stakeholders in a secure environment. Thus, blockchain integration improves the traceability of items and reduces the waiting time, thereby improving supply chain operational efficiency. All the information is stored in a centralized immutable digital ledger with varying levels of access for different parties. If implemented on a global scale, this network has the potential to improve the overall supply chain operations of the automotive industry.

Implementation of blockchain can potentially overcome the issues faced by the automotive industry such as traceability issues at various nodes of an automotive supply chain. Differential pricing can be availed by using a permission ledger for confidential transactions between parties. Consensus between multiple parties is maintained through Smart Contracts which helps to intimate required information between different mediating parties [55]. The disruptions are rather promoted by blockchain as the digital ledger is free of geographical constraints such as natural disasters and socio-economic issues. Furthermore, the smart contract only lets out payments once both parties are satisfied, and the risk of fraud is mitigated by using nodes for checking delivery status. Due to using doubly signed smart contracts, no financial fraud by middlemen can occur in the system. Table 4 describes how the proposed solution helps in solving common organizational-level supply chain problems.

Table 4. Solution description.

Problems	How the Proposed Solution Solves Them
Tracking supply-chain history of each part	Every party has access to their respective order status where they can track the product in real-time.
Differential Pricing	Permission ledger for confidential transactions between parties
Numerous parties involved	Consensus between multiple parties is maintained through Smart Contracts
Quality and Compliance issues	Smart Contract stores money while all solutions are checked and tested
Inevitable disruptions	The digital ledger is free of geographical constraints like natural disasters, socio-economic issues
Procuring replacements for defective pieces	The smart contract only lets out payments once both parties satisfied
Centralization	The risk of fraud is mitigated by using nodes for checking delivery status
Fraud by Middlemen	Due to using doubly signed smart contracts, no financial fraud by middlemen can occur in the system

5.1. Theoretical Implications

The focus on the blockchain is seen more in areas of financial services, food, consumer services, and healthcare [56–59]. This article provides multiple theoretical contributions to the existing literature on the blockchain. The major challenges of the automotive supply chain such as traceability issues and waiting time issues are not focused on earlier. First, the study is made based on an analysis of a contemporary organization that looks for digital technologies to overcome traceability and waiting time issues. Secondly, a blockchain architecture is proposed for an automobile supply chain that can be used by other industries facing similar problems. Further, a simulation-based study is conducted for integrating blockchain that was not focused on earlier. This study is intended to contribute to the blockchain literature with the objective of how hyper ledger fabric-based blockchain can help the automotive industry to reduce daily operational costs, mean waiting time, and improve responsiveness. In this context, the article aims to contribute to Industry 4.0 literature as well.

5.2. Practical Implications

The study offers practical implications for supply chain stakeholders such as OEM, 3PL/Freight Forwarder, Dealer, and Customer. The emerging technology blockchain in industry 4.0 aids organizations to overcome operational challenges by ensuring secured and authentic information. Typically, manufacturers can own and manage the blockchain network. Transparency is achieved by various nodes acting as individual participants of the supply chain. In some cases, the blockchain can be exclusive to the company itself or the blockchain network can be distributed among various stakeholders and company partners. In this study, a hyper ledger fabric-based private blockchain is employed wherein certain parts of the information can be accessed by certain entities considering the complexity involved in sharing information.

When compared to the data management frameworks of conventional traceability tools such as RFID tagging and E-Kanban; blockchain has the potential to overcome other tool's limitations and improve supply chain operations. A blockchain will improve supply chain operations by enabling an interconnected network of the supply chain. There are several studies done with public and consortium blockchain. For public blockchain, the immutability is good, but the efficiency is low. It takes much time to transmit the data when the volume increases [60,72,73]. For consortium blockchain, the efficiency is good, but it is partially decentralized which makes it difficult to control by primary supply chain members [74–76]. In this article, a new blockchain architecture is proposed based on private blockchain where efficiency is high. Moreover, it is entirely centralized so that primary supply chain members can control it. Hyperledger Fabric blockchain is a private blockchain that can be implemented in a wide range of automotive industries. It maximizes confidentiality and scalability. It uses Smart Contracts where supply chain members can manage their transactions and collaborate with the supply chain network. Table 5 shows the limitations faced by different stakeholders in a global supply chain without blockchain implementation and how the same can be overcome with our proposed blockchain architecture.

Table 5. Comparison of supply chain with and without our proposed blockchain architecture.

Stakeholders	Without Blockchain	With Blockchain
OEM Adds value to the raw materials into other consumables and finally the end product	Has a limited ability to control as well as to verify the flows coming from its suppliers? (e.g., proper compliance to standards and requirements)	Benefits from an integrated and distributed ledger that enables them to control the inputs and keep track of their production
3PL/Freight Forwarder Responsible for transporting materials and products between stakeholders	Reliable but the one-sided tracking system Limited certification ability and complex tracking (e.g., heat or pressure variations) Difficulty to certify a code of conduct	Shared information system The client can benefit from a distributed and certified system The client can make sure his goods are transported in the right conditions and timing
Dealer The link between the OEM and the end consumer	Difficulty to certify the origin and path of the goods bought and sold	Can easily check the origin of the goods and their transformation path on the blockchain. With sealed IoT devices put on the goods, the broker can check and prove their authenticity and provenance
Person/Customer The end consumer of the product	Difficulty to verify the total compliance, origin, and composition of the goods to be bought	Has a full view of the goods bought (i.e., provenance, transformation process, transportation) directly on the blockchain

Based on the discussion with the case organization, they are keen on expanding the pilot study in other plants as well considering the value the blockchain features offer in terms of improving traceability and reducing waiting times thereby increasing the supply chain operational efficiency. In addition, supply chain members believe that the resources can be effectively managed with technology and are keen on investing in new technologies to improve operational efficiency thereby savings costs. The outcome from the blockchain technology study will gain practitioners' attention who are looking for a technology-based solution that impacts their organization and connected supply chain members. The improved traceability in inventory and reduced waiting time and costs impact the financial revenues for the organization. Thus, the management is keen on implementing blockchain based on the simulation study. In addition, transforming into a digital organization gives them a competitive advantage in the automotive sector that improves their reputation, thereby increasing its customers.

6. Conclusions

The need for blockchain integration is being endorsed by some organizations, however, this view is still not shared by many companies. This study aimed at identifying traceability issues at various nodes of an automotive supply chain. A new blockchain architecture for the automotive supply chain is proposed. The results were obtained after optimizing the current supply chain by considering blockchain integration. The results seemed to be promising, as there was improved communication between the partners indicated results in improved traceability in inventory that improves IQR and reduction of mean waiting time thereby improving overall supply chain efficiency. Mean daily costs of the whole supply chain were also seen to be reduced and the traceability on waiting time, inventory can be improved with the help of hyper ledger-based blockchain. The blockchain embedded supply chain is more capable in eliminating the risks and uncertainties associated with the automotive supply chain. These are the benefits of adopting blockchain technology in the automotive supply chain.

Five blockchain experts were further consulted to compare the supply chain and blockchain functionalities identified. Here, traceability was found to be the second most important factor and its blockchain readiness value was significantly higher than supply chain readiness which suggests that the current traceability system can be effective if blockchain is implemented. Blockchain offers a peer-to-peer network with cost-effective transaction and data security, thus making its integration with newer technologies all the more desirable. This would inherently assist any organization in making its supply chain cost-effective in the long run.

In conclusion, this study offered a new perspective on implementing blockchain and further digitizing the organization as reviewed by industry and business experts. However, one limitation of this study is that it offers insight only into the automotive sector. Moreover, the industry experts consulted were also predominantly from a manufacturing background. Further research may include the study of how cloud services may accelerate the integration, blockchain's integration capability with other tools, and the system's isolation from inevitable disruptions. The unique contributions of the study are

- A framework for blockchain is developed for the automotive supply chain where stakeholders get visibility into goods movement, inventory status, and waiting time.
- A hyper ledger fabric-based blockchain architecture is designed for an automotive supply chain to track the ownership transfers in inbound and outbound logistics, order readiness to fulfill the demand.

The implementation process for blockchain-enabled supply chain architecture mostly revolves around capital issues, network infrastructure issues, and legal aspects of implementation. Except for bitcoin, other use cases for blockchain have not been explored to an adequate level. For example, there is a limited amount of work done to address supply chain-blockchain integration keeping global context in mind. There is an opportunity to study the effectiveness of blockchain on a variety of supply chain operational parameters. The four core values of blockchain are still being closely researched and more functionalities may also be added in the future. Many governments around the globe have not made any laws or regulations for blockchain implementation. The decentralized database will provide more trust for regulators, and it improves the visibility among supply chain members. Researchers can focus on the security and legal aspects of blockchain-enabled supply chain implementation. With disruptive technologies coming into play and technical advancements being conducted with Industry 4.0, a comprehensive study can be performed to address any possible loopholes and issues in the blockchain network. The hyper ledger architecture developed in this study can also be expanded for advanced features such as geofencing and live location sharing with the help of sensors and IoT integration. It is evident that blockchain is evolving and thorough analysis on limitations also needs to be studied before implementation. This may help the practitioners and academia for further research on it.

Author Contributions: N.A.: Data curation. M.E.: Writing—Original draft preparation, Software. A.K.: Validation, Review, and editing. V.K.E.K.: Visualization, Investigation. S.P.N.: Formal analysis, Validation. Y.K.: Methodology. J.K.: Conceptualization, Project administration, Supervision. All authors have read and agreed to the published version of the manuscript.

Funding: This research received no external funding.

Institutional Review Board Statement: Not applicable.

Informed Consent Statement: Informed consent was obtained from the respondents of the survey.

Data Availability Statement: The data will be made available on request from the corresponding author.

Conflicts of Interest: The authors declare no conflict of interest.

References

1. Building Resilience in Global Supply Chains (no date). Available online: https://docs.wbcsd.org/2015/12/building-resilience-in-global-supply-chains.pdf (accessed on 27 March 2019).
2. Longo, F.; Nicoletti, L.; Padovano, A.; D'Atri, G.; Forte, M. Blockchain-enabled supply chain: An experimental study. *Comput. Ind. Eng.* **2019**, *136*, 57–69. [CrossRef]
3. Ma, F.; Xue, H.; Yuen, K.F.; Sun, Q.; Zhao, S.; Zhang, Y.; Huang, K. Assessing the Vulnerability of Logistics Service Supply Chain Based on Complex Network. *Sustainability* **2020**, *12*, 1991. [CrossRef]
4. Brody, P. How blockchain is revolutionizing supply chain management. *Digit. Mag.* **2018**, 1–7.
5. Pirola, F.; Boucher, X.; Wiesner, S.; Pezzotta, G. Digital technologies in product-service systems: A literature review and a research agenda. *Comput. Ind.* **2020**, *123*, 103301. [CrossRef]
6. Litke, A.; Anagnostopoulos, D.; Varvarigou, T. Blockchains for Supply Chain Management: Architectural Elements and Challenges Towards a Global Scale Deployment. *Logistics* **2019**, *3*, 5. [CrossRef]
7. Westerkamp, M.; Victor, F.; Küpper, A. Tracing manufacturing processes using blockchain-based token compositions. *Digit. Commun. Netw.* **2019**, *6*, 167–176. [CrossRef]
8. Khan, S.; Haleem, A.; Khan, M.I.; Abidi, M.H.; Al-Ahmari, A. Implementing Traceability Systems in Specific Supply Chain Management (SCM) through Critical Success Factors (CSFs). *Sustainability* **2018**, *10*, 204. [CrossRef]
9. Sarkar, M.; Seo, Y.W. Renewable energy supply chain management with flexibility and automation in a production system. *J. Clean. Prod.* **2021**, *324*, 129149. [CrossRef]
10. Casino, F.; Kanakaris, V.; Dasaklis, T.K.; Moschuris, S.; Stachtiaris, S.; Pagoni, M.; Rachaniotis, N.P. Block-chain-based food supply chain traceability: A case study in the dairy sector. *Int. J. Prod. Res.* **2021**, *59*, 5758–5770.
11. Fan, Y.; Wu, S.; Lu, Y.; Zhao, Y. Study on the effect of the environmental protection industry and investment for the national economy: An input-output perspective. *J. Clean. Prod.* **2019**, *227*, 1093–1106. [CrossRef]
12. Ning, F.; Shi, Y.; Cai, M.; Xu, W.; Zhang, X. Manufacturing cost estimation based on the machining process and deep-learning method. *J. Manuf. Syst.* **2020**, *56*, 11–22. [CrossRef]
13. Hasan, H.R.; Salah, K.; Jayaraman, R.; Ahmad, R.W.; Yaqoob, I.; Omar, M. Blockchain-Based Solution for the Traceability of Spare Parts in Manufacturing. *IEEE Access* **2020**, *8*, 100308–100322. [CrossRef]
14. Shao, L.; Jin, S. Resilience assessment of the lithium supply chain in China under impact of new energy vehicles and supply interruption. *J. Clean. Prod.* **2019**, *252*, 119624. [CrossRef]
15. Cognex. Cognex Expert Guide: Traceability for the Automotive Industry Cognex. 2017. Available online: https://www.cognex.com/resources/white-papers-articles/expert-guide-traceability-for-the-automotive-industry-9334 (accessed on 17 January 2019).

16. Dai, B.; Nu, Y.; Xie, X.; Li, J. Interactions of traceability and reliability optimization in a competitive supply chain with product recall. *Eur. J. Oper. Res.* **2021**, *290*, 116–131. [CrossRef]
17. Alfian, G.; Syafrudin, M.; Rhee, J. Real-time monitoring system using smartphone-based sensors and nosql database for perishable supply chain. *Sustainability* **2017**, *9*, 2073. [CrossRef]
18. Dallasega, P.; Rauch, E.; Linder, C. Industry 4.0 as an enabler of proximity for construction supply chains: A sys-tematic literature review. *Comput. Ind.* **2018**, *99*, 205–225. [CrossRef]
19. Abu Zwaida, T.; Pham, C.; Beauregard, Y. Optimization of Inventory Management to Prevent Drug Shortages in the Hospital Supply Chain. *Appl. Sci.* **2021**, *11*, 2726. [CrossRef]
20. Pourhejazy, P. Destruction Decisions for Managing Excess Inventory in E-Commerce Logistics. *Sustainability* **2020**, *12*, 8365. [CrossRef]
21. Kamble, S.S.; Gunasekaran, A.; Sharma, R. Modeling the blockchain enabled traceability in agriculture supply chain. *Int. J. Inf. Manag.* **2019**, *52*, 101967. [CrossRef]
22. Korpela, K.; Hallikas, J.; Dahlberg, T. Digital supply chain transformation toward blockchain integration. In Proceedings of the 50th Hawaii International Conference on System Sciences, Waikoloa, HI, USA, 4–7 September 2017.
23. Pan, X.; Pan, X.; Song, M.; Ai, B.; Ming, Y. Blockchain technology and enterprise operational capabilities: An empirical test. *Int. J. Inf. Manag.* **2019**, *52*, 101946. [CrossRef]
24. Di Vaio, A.; Varriale, L. Management innovation for environmental sustainability in seaports: Managerial accounting instruments and training for competitive green ports beyond the regulations. *Sustainability* **2018**, *10*, 783. [CrossRef]
25. Petri, I.; Barati, M.; Rezgui, Y.; Rana, O.F. Blockchain for energy sharing and trading in distributed prosumer communities. *Comput. Ind.* **2020**, *123*, 103282. [CrossRef]
26. Tandon, A.; Dhir, A.; Islam, A.N.; Mäntymäki, M. Blockchain in healthcare: A systematic literature review, synthesizing framework and future research agenda. *Comput. Ind.* **2020**, *122*, 103290. [CrossRef]
27. Bonilla, S.H.; Silva, H.R.; Terra da Silva, M.; Franco Gonçalves, R.; Sacomano, J.B. Industry 4.0 and Sustainability Implications: A Scenario-Based Analysis of the Impacts and Challenges. *Sustainability* **2018**, *10*, 3740. [CrossRef]
28. Di Vaio, A.; Varriale, L. Blockchain technology in supply chain management for sustainable performance: Evidence from the airport industry. *Int. J. Inf. Manag.* **2019**, *52*, 102014. [CrossRef]
29. Junaid, M.; Xue, Y.; Syed, M.W.; Li, J.Z.; Ziaullah, M. A neutrosophic ahp and topsis framework for supply chain risk assessment in automotive industry of Pakistan. *Sustainability* **2020**, *12*, 154. [CrossRef]
30. González-Benito, J.; Lannelongue, G.; Alfaro-Tanco, J.A. Study of supply-chain management in the automotive industry: A bibliometric analysis. *Int. J. Prod. Res.* **2013**, *51*, 3849–3863. [CrossRef]
31. Maro, S.; Steghöfer, J.-P.; Staron, M. Software traceability in the automotive domain: Challenges and solutions. *J. Syst. Softw.* **2018**, *141*, 85–110. [CrossRef]
32. Papetti, A.; Marconi, M.; Rossi, M.; Germani, M. Web-based platform for eco-sustainable supply chain management. *Sustain. Prod. Consum.* **2018**, *17*, 215–228. [CrossRef]
33. Ferriols, F.J.; Mula, J.; Díaz-Madroñero, M. Supply chain management as the company engine in automotive man-ufacturing. *IFAC Proc. Vol.* **2013**, *46*, 682–687. [CrossRef]
34. Hastig, G.M.; Sodhi, M.S. Blockchain for Supply Chain Traceability: Business Requirements and Critical Success Factors. *Prod. Oper. Manag.* **2019**, *29*, 935–954. [CrossRef]
35. Felsberger, A.; Qaiser, F.H.; Choudhary, A.; Reiner, G. The impact of Industry 4.0 on the reconciliation of dynamic capabilities: Evidence from the European manufacturing industries. *Prod. Plan. Control.* **2020**, *1*, 1–24. [CrossRef]
36. Benatia, M.A.; Baudry, D.; Louis, A. Detecting counterfeit products by means of frequent pattern mining. *J. Ambient. Intell. Humaniz. Comput.* **2020**, *1*, 1–10. [CrossRef]
37. Liu, Z.; Li, Z. A blockchain-based framework of cross-border e-commerce supply chain. *Int. J. Inf. Manag.* **2019**, *52*, 102059. [CrossRef]
38. Danzon, P.M.; Towse, A. Differential Pricing for Pharmaceuticals: Reconciling Access, R&D and Patents. *Int. J. Health Care Finance Econ.* **2003**, *3*, 183–205. [CrossRef]
39. Tsao, Y.-C.; Vu, T.-L. Power supply chain network design problem for smart grid considering differential pricing and buy-back policies. *Energy Econ.* **2019**, *81*, 493–502. [CrossRef]
40. Liu, Z.; Gao, R.; Zhou, C.; Ma, N. Two-period pricing and strategy choice for a supply chain with dual uncertain information under different profit risk levels. *Comput. Ind. Eng.* **2019**, *136*, 173–186. [CrossRef]
41. Queiroz, M.M.; Wamba, S.F. Blockchain adoption challenges in supply chain: An empirical investigation of the main drivers in India and the USA. *Int. J. Inf. Manag.* **2018**, *46*, 70–82. [CrossRef]
42. Li, Z.; Liu, L.; Barenji, A.V.; Wang, W. Cloud-based manufacturing blockchain: Secure knowledge sharing for in-jection mould redesign. *Procedia CIRP* **2018**, *72*, 961–966. [CrossRef]
43. Li, Z.; Barenji, A.V.; Huang, G.Q. Toward a blockchain cloud manufacturing system as a peer to peer distributed network platform. *Robot. Comput. Manuf.* **2018**, *54*, 133–144. [CrossRef]
44. Muzammal, M.; Qu, Q.; Nasrulin, B. Renovating blockchain with distributed databases: An open source system. *Futur. Gener. Comput. Syst.* **2018**, *90*, 105–117. [CrossRef]
45. Banerjee, A. Blockchain Technology: Supply Chain Insights from ERP. *Adv. Comput.* **2018**, 69–98. [CrossRef]

46. Atlam, H.F.; Wills, G.B. Technical aspects of blockchain and IoT. In *Advances in Computers*; Academic Press Inc.: Cambridge, MA, USA, 2019; Volume 115, pp. 1–39. ISBN 9780128171899.
47. Min, H. Blockchain technology for enhancing supply chain resilience. *Bus. Horiz.* **2019**, *62*, 35–45. [CrossRef]
48. Xu, X.; Lu, Q.; Liu, Y.; Zhu, L.; Yao, H.; Vasilakos, A.V. Designing blockchain-based applications a case study for imported product traceability. *Futur. Gener. Comput. Syst.* **2018**, *92*, 399–406. [CrossRef]
49. Casado-Vara, R.; Prieto, J.; de la Prieta, F.; Corchado, J.M. How blockchain improves the supply chain: Case study alimentary supply chain. *Procedia Comput. Sci.* **2018**, *134*, 393–398. [CrossRef]
50. Costa, F.; do Sameiro Carvalho, M.; Fernandes, J.M.; Alves, A.C.; Silva, P. Improving visibility using RFID–the case of a company in the automotive sector. *Procedia Manuf.* **2017**, *13*, 1261–1268. [CrossRef]
51. Kamble, S.S.; Gunasekaran, A.; Subramanian, N.; Ghadge, A.; Belhadi, A.; Venkatesh, M. Blockchain technology's impact on supply chain integration and sustainable supply chain performance: Evidence from the automotive industry. *Ann. Oper. Res.* **2021**, 1–26. [CrossRef]
52. Khanfar, A.; Iranmanesh, M.; Ghobakhloo, M.; Senali, M.; Fathi, M. Applications of Blockchain Technology in Sustainable Manufacturing and Supply Chain Management: A Systematic Review. *Sustainability* **2021**, *13*, 7870. [CrossRef]
53. Rejeb, A.; Rejeb, K.; Simske, S.; Treiblmaier, H. Blockchain Technologies in Logistics and Supply Chain Management: A Bibliometric Review. *Logistics* **2021**, *5*, 72. [CrossRef]
54. Azzi, R.; Chamoun, R.K.; Sokhn, M. The power of a blockchain-based supply chain. *Comput. Ind. Eng.* **2019**, *135*, 582–592. [CrossRef]
55. Zheng, Z.; Xie, S.; Dai, H.-N.; Chen, W.; Chen, X.; Weng, J.; Imran, M. An overview on smart contracts: Challenges, advances and platforms. *Future Gener. Comput. Syst.* **2019**, *105*, 475–491. [CrossRef]
56. Beck, R.; Avital, M.; Rossi, M.; Thatcher, J. Blockchain Technology in Business and Information Systems Research. *Bus. Inf. Syst. Eng.* **2017**, *59*, 381–384. [CrossRef]
57. Treiblmaier, H. Toward More Rigorous Blockchain Research: Recommendations for Writing Blockchain Case Studies. *Front. Blockchain* **2019**, *2*, 1–31. [CrossRef]
58. Ali, O.; Ally, M.; Clutterbuck; Dwivedi, Y. The state of play of blockchain technology in the financial services sector: A systematic literature review. *Int. J. Inf. Manag.* **2020**, *54*, 102199. [CrossRef]
59. Soltanisehat, L.; Alizadeh, R.; Hao, H.; Choo, K.-K.R. Technical, Temporal, and Spatial Research Challenges and Opportunities in Blockchain-Based Healthcare: A Systematic Literature Review. *IEEE Trans. Eng. Manag.* **2020**, *1*, 1–16. [CrossRef]
60. Zheng, Z.; Xie, S.; Dai, H.; Chen, X.; Wang, H. An overview of blockchain technology: Architecture, consensus, and future trends. In Proceedings of the 2017 IEEE international congress on big data (BigData congress), Honolulu, HI, USA, 25–30 June 2017; IEEE: Piscataway, NJ, USA, 2017; pp. 557–564.
61. Özceylan, E.; Demirel, N.; Çetinkaya, C.; Demirel, E. A closed-loop supply chain network design for automotive industry in Turkey. *Comput. Ind. Eng.* **2017**, *113*, 727–745. [CrossRef]
62. Sher, M.M.; Kim, S.-L.; Banerjee, A.; Paz, M.T. A supply chain coordination mechanism for common items subject to failure in the electronics, defense, and medical industries. *Int. J. Prod. Econ.* **2018**, *203*, 164–173. [CrossRef]
63. Al-Aomar, R.; Hussain, M. An assessment of adopting lean techniques in the construct of hotel supply chain. *Tour. Manag.* **2018**, *69*, 553–565. [CrossRef]
64. Kalverkamp, M.; Young, S.B. In support of open-loop supply chains: Expanding the scope of environmental sus-tainability in reverse supply chains. *J. Clean. Prod.* **2019**, *214*, 573–582. [CrossRef]
65. Masoud, S.A.; Mason, S.J. Integrated cost optimization in a two-stage, automotive supply chain. *Comput. Oper. Res.* **2016**, *67*, 1–11. [CrossRef]
66. Borgstedt, P.; Neyer, B.; Schewe, G. Paving the road to electric vehicles–A patent analysis of the automotive supply industry. *J. Clean. Prod.* **2017**, *167*, 75–87. [CrossRef]
67. Mathivathanan, D.; Kannan, D.; Haq, A.N. Sustainable supply chain management practices in Indian automotive industry: A multi-stakeholder view. *Resour. Conserv. Recycl.* **2018**, *128*, 284–305. [CrossRef]
68. Ambe, I.M.; Badenhorst-Weiss, J.A. An automotive supply chain model for a demand-driven environment. *J. Transp. Supply Chain Manag.* **2011**, *5*, 1–22. [CrossRef]
69. Sundarakani, B.; de Souza, R.; Goh, M. A Case Study of Singapore's Automotive Supply Chain: Preliminary Analysis and Findings. In *Cases on Supply Chain and Distribution Management: Issues and Principles*; IGI Global: Hershey, PA, USA, 2012; pp. 200–254.
70. Bhattacharya, S.; Mukhopadhyay, D.; Giri, S. Supply chain management in Indian automotive industry: Complexi-ties, challenges and way ahead. *Int. J. Manag. Value Supply Chain.* **2014**, *5*, 49. [CrossRef]
71. Teucke, M.; Sommerfeld, D.; Freitag, M. Sharing Sensor Based Quality Data in Automotive Supply Chain Processes. *IFAC-PapersOnLine* **2018**, *51*, 770–775. [CrossRef]
72. Kaynak, B.; Kaynak, S.; Uygun, O. Cloud Manufacturing Architecture Based on Public Blockchain Technology. *IEEE Access* **2019**, *8*, 2163–2177. [CrossRef]
73. Yang, J.; Paudel, A.; Gooi, H.B. Compensation for Power Loss by a Proof-of-Stake Consortium Blockchain Microgrid. *IEEE Trans. Ind. Informatics* **2020**, *17*, 3253–3262. [CrossRef]

74. Yang, R.; Wakefield, R.; Lyu, S.; Jayasuriya, S.; Han, F.; Yi, X.; Chen, S. Public and private blockchain in con-struction business process and information integration. *Autom. Constr.* **2020**, *118*, 103276. [CrossRef]
75. Zhang, S.; Rong, J.; Wang, B. A privacy protection scheme of smart meter for decentralized smart home environment based on consortium blockchain. *Int. J. Electr. Power Energy Syst.* **2020**, *121*, 106140. [CrossRef]
76. Yuan, K.; Wu, G.; Dong, H.; He, B.; Wang, D. Differential Pricing and Emission Reduction in Remanufacturing Supply Chains with Dual-Sale Channels under CCT-Mechanism. *Sustainability* **2020**, *12*, 8150. [CrossRef]

sustainability

MDPI

Article

Blockchain-Based Model to Improve the Performance of the Next-Generation Digital Supply Chain

Sumit Kumar Rana [1], Hee-Cheol Kim [2], Subhendu Kumar Pani [3], Sanjeev Kumar Rana [1], Moon-Il Joo [4,*], Arun Kumar Rana [5] and Satyabrata Aich [6,*]

1 Computer Science and Engineering Department, Maharishi Markandeshwar (Deemed to be University), Ambala 133207, India; sumitrana.cse@gmail.com (S.K.R.); dr.sanjeevrana@mmumullana.org (S.K.R.)
2 College of AI Convergence/Institute of Digital Anti-Aging Healthcare/u-AHRC, Inje University, Gimhae 50834, Korea; heeki@inje.ac.kr
3 Department of Computer Applications, Krupajal Computer Academy, Bhubaneswar 751002, India; pani.subhendu@gmail.com
4 Institute of Digital Anti-Aging Healthcare, Inje University, Gimhae 50834, Korea
5 Department of Electronics and Communication Engineering, Panipat Institute of Engineering and Technology, Samalkha, Panipat 132102, India; ranaarun1@gmail.com
6 Wellmatix Corporation Limited, Changwon 51395, Korea
* Correspondence: joomi@inje.ac.kr (M.-I.J.); satyabrataaich@gmail.com (S.A.); Tel.: +82-55-320-4027 (S.A.)

Abstract: In the era of the fourth industrial revolution, all aspects of the industrial domain are being affected by emerging technologies. Digitalization of every process is taking place or under process. One of the most important components common to every domain is the supply chain process. Organizations employ a digital supply chain to track the delivery of their products or materials. The digital supply chain is still suffering from a few issues such as no provenance, less transparency, and a trust issue. Blockchain technology, one of the emerging technologies, can be integrated with the supply chain to deal with the existing issues and to improve its performance. In this paper, a model is proposed to integrate blockchain technology with the supply chain to improve performance. The proposed model uses the combination of the Ethereum blockchain and the interplanetary file system to maintain the traceability, transparency, and trustworthiness of the supply chain.

Keywords: blockchain; supply chain; digitalization; Ethereum blockchain; IPFS

Citation: Rana, S.K.; Kim, H.-C.; Pani, S.K.; Rana, S.K.; Joo, M.-I.; Rana, A.K.; Aich, S. Blockchain-Based Model to Improve the Performance of the Next-Generation Digital Supply Chain. *Sustainability* **2021**, *13*, 10008. https://doi.org/10.3390/su131810008

Academic Editor: Mark A. Bonn

Received: 17 July 2021
Accepted: 31 August 2021
Published: 7 September 2021

Publisher's Note: MDPI stays neutral with regard to jurisdictional claims in published maps and institutional affiliations.

Copyright: © 2021 by the authors. Licensee MDPI, Basel, Switzerland. This article is an open access article distributed under the terms and conditions of the Creative Commons Attribution (CC BY) license (https://creativecommons.org/licenses/by/4.0/).

1. Introduction

A supply chain is a series of steps in the production, transportation, storage, or distribution of a product [1]. Each stage may be handled by a single company, a group of companies, or a group of stakeholders. The global economy relies heavily on supply chains [2]. According to the International Trade Administration, supply chain transactions account for more than 76 percent of global trade [3]. To reduce production costs, large corporations outsource assembly lines to low-cost regions. The stages of the supply chain have been further divided, and an increasing number of affiliates are now in charge of them. Throughout the stages, supply chains have become more global, complex, and interdependent [4]. The survival of any organization depends on how good the synchronization is between different businesses functions involved in the process. One of the most important components of the business process is the supply chain. The supply chain is an integrated chain of different business functions, namely acquiring raw materials, delivery of products or services, etc. that are required to run a business. It also connects with all the stakeholders involved in the business. Effective supply chain management is a source of competitive advantage across various industries, including food, healthcare, IT, etc. [5]. If we look at recent history, several incidents related to the safety of agriproducts have occurred. A few examples are "poisonous ginger" that happened in China, infection in Hami melon due to listeria in the United States, and the epidemic of *E. coli* in Germany. Because of these incidents, countries are more attentive towards the traceability of agriproducts and regularly bring laws and regulations to the supply chain process. A traceability procedure must be attached with every supply chain so that in case of any outbreak, recall can be executed in an efficient manner and information can be passed to respective consumers. A traditional supply chain ecosystem is shown in Figure 1. In this type of approach, all the data related to all the processes is stored on a central database. This central database is taken care of by an administrator. Some limitations of this approach are mentioned in the following. As this uses a centralized server, if the central database server goes down, then the complete system will be down. This server is maintained by an administrator, if he is dishonest, he can manipulate the data without the knowledge of the stakeholders. These changes cannot be traced. So, this centralized approach also lacks traceability and transparency.

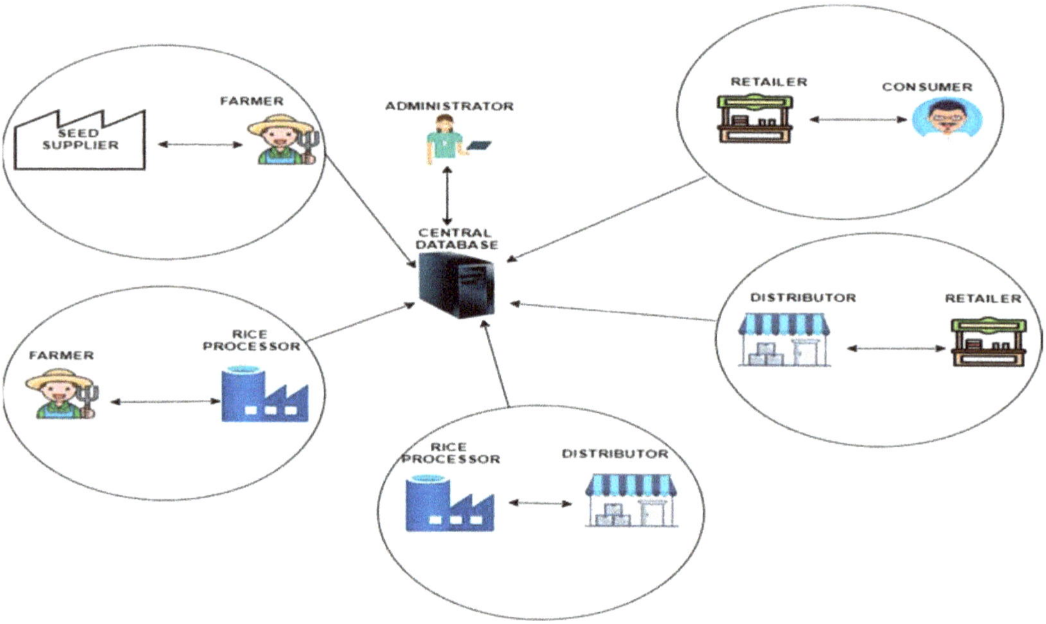

Figure 1. Traditional Supply Chain Ecosystem.

Major issues with the traditional supply chain ecosystem are traceability of the products, transparency for the stakeholders, trust in the collaborative system, and security from unauthorized modification of data. As there are multiple intermediaries in the traditional approach, trust and performance issues are common. There are multiple entities in the supply chain ecosystem such as farmers, distributors, retailers, etc. These can be in multiples and in different regions which makes it a complex network of entities. So, if there is an outbreak of any food-related contamination then traceability of the products is very difficult and time-consuming. Functional impact, social impact, economic impact, and the issues arising during the integration of emerging technologies with the supply chain ecosystem need to be investigated [6]. Second, the traditional supply chain ecosystem is completely centralized. This centralization brings trust issues in the collaborative environment where multiple entities are involved. In this centralized approach, data can be easily manipulated without the knowledge of other stakeholders. In the case of the food supply chain which delivers daily need products, traceability is of main concern because any carelessness may put the life or health of people at stake. Trust of the consumer should be the utmost priority for companies because trust issues in the supply chain process can cause significant losses.

This process must be transparent to all the stakeholders of the process. So, to regain the trust of the consumer there must be some way to verify the origin of the food and to stop fraud [7].

These issues can be resolved if user data can be made accessible to all whereas alteration of data is not possible by anyone. To overcome these issues an emerging technology can be employed with the supply chain. Blockchain is the technology that can be used to remove the issues and improve the performance of the supply chain. It uses distributed ledger technology that has some features which can help in dealing with issues of the supply chain process. It is a distributed and immutable ledger that provides a trustable record that cannot be manipulated or tampered with by any entity. Transparency in the food supply chain and use of emerging technologies can improve the conventional supply chain ecosystem [8]. Data security and reliability, information interconnection and intercommunication, real-time sharing of hazardous-material information, and dynamic and credible whole-process tracing are missing in traditional systems [9]. Researchers are regularly working to provide a better understanding of blockchain innovation by presenting some key features and guidelines for accelerating blockchain adoption in industry [10].

This paper proposes a model supported by blockchain and smart contracts to trace the flow of data in the supply chain process of rice from its origin to the consumer. This model eliminates the need for intermediaries for different information sharing among the various stakeholders and advances the level of transparency, trust, and security. Different entities upload environmental data, rice growth data, buy and sell order data into the Interplanetary File System at the same time. The IPFS hash of the file is saved in smart contracts, which helps to solve the blockchain storage problem.

Motivation and Contribution

The motivation behind this research is the emergence of blockchain technology and its application in the supply chain ecosystem. Blockchain can provide better security, improved trust level, and avoid unauthorized modification due to the immutability feature. The main contributions of this paper are:

- Discussion of the recent publications on merging blockchain technology with the supply chain ecosystem.
- Investigation of the issues of the traditional supply chain and how these issues can be resolved by amalgamating blockchain technology.
- Presentation of a model that uses blockchain technology to improve the performance of the supply chain ecosystem.
- Proposition of a model that ensures transparency, traceability, and trustworthiness.

The remaining parts of the paper are structured as follows: In Section 2, research work of various authors related to blockchain application in the supply chain domain is explored. In Section 3, blockchain technology and its advantages in the supply chain domain are discussed. Section 4 presents a blockchain-supported model for the supply chain ecosystem. Section 5 concludes the paper and discusses future research directions.

2. Related Study

A. Shahid et al. presented a complete solution for the blockchain-based Agriculture and Food (Agri-Food) supply chain in our proposed solution. It makes use of the Ethereum blockchain network's key features of blockchain and smart contracts. Although blockchain ensures the immutability of data and records in the network, it still falls short of solving some major issues in supply chain management, such as the credibility of the parties involved, trading process accountability, and product traceability. Therefore, there is a need for a reliable system that ensures traceability, trust, and delivery mechanism in the Agri-Food supply chain. In the proposed system, all transactions are written to blockchain which ultimately uploads the data to the Interplanetary File Storage System (IPFS). The storage system returns a hash of the data which is stored on the blockchain and ensures an efficient, secure, and reliable solution [11].

A. Musamih et al. presented an Ethereum blockchain-based approach for efficient product traceability in the healthcare supply chain that uses smart contracts and decentralized off-chain storage. The smart contract ensures data provenance, eliminates the need for intermediaries, and provides all stakeholders with a secure, immutable transaction history. We present the system architecture as well as the detailed algorithms that govern our proposed solution's working principles. We test and validate the system, as well as present a cost and security analysis, to assess its effectiveness in improving traceability within pharmaceutical supply chains [12].

A. Kamilaris et al. looked at the impact of blockchain technology on agriculture and the food supply chain. They presented current projects and initiatives and discussed overall implications, challenges, and potential, all the while keeping a critical eye on the projects' maturity. Their findings showed that blockchain is a promising technology for a transparent food supply chain, with numerous ongoing initiatives in a variety of food products and food-related issues, but there are still many barriers and challenges that prevent it from gaining broader adoption among farmers and systems. Technical aspects, education, policies, and regulatory frameworks are all part of these challenges [13].

In business practice, there are numerous applications of BCT in supply chain management (SCM), and there is growing interest in this topic among academics. G. Blossey et al. combined these two perspectives on BCT in SCM to summarize the current state of the art and to identify areas for future research. A comprehensive framework of BCT using case clusters in SCM has been developed for this purpose, based on the unique features of BCT. The framework was used to examine 53 BCT applications in SCM that were discovered through a systematic literature review and a secondary dataset of blockchain-driven SCM innovations [14].

The potential application of blockchain technology and smart contracts to supply chain management was critically examined by S. Saberi et al. Government, community, and consumer pressures on local and global governments, communities, and consumers to meet sustainability goals prompted us to investigate how blockchain can address and aid supply chain sustainability. Part of this critical examination was to see how blockchains, a potentially disruptive technology still in its early stages of development, can overcome a variety of obstacles. Inter-organizational, intra-organizational, technical, and external barriers to blockchain technology adoption were discussed [15].

P. Gonczol et al. reviewed academic research and implementations of distributed ledgers on supply chains. They summarized the benefits and challenges of distributed supply chain organization and management, as well as the current state of research on the topic. They discussed the technical characteristics and maturity of various industrial projects, focusing on industrial practices and use cases. Their goal was to evaluate the utility of blockchains in the supply chain domain and to lay the groundwork for practitioners and researchers to focus their future projects on improving the technology and its applications [16].

A blockchain-IoT-based food traceability system (BIFTS) was proposed by Y.P.Tsang et al. The authors integrated the novel deployment of blockchain, IoT technology, and fuzzy logic into a total traceability shelf-life management system for perishable food management. Lightweight and vaporized characteristics were deployed in the blockchain to address the needs for food traceability, while an integrated consensus mechanism that considered shipment transit time, stakeholder assessment, and shipment volume was developed. The blockchain data flow was then synchronized with the deployment of IoT technologies at the level of traceable resource units [17].

K.Salah et al. proposed a method for tracking and tracing soybeans across the agricultural supply chain that used the Ethereum blockchain and smart contracts to efficiently perform business transactions. Their proposed solution eliminates the need for a trusted centralized authority, intermediaries, and transaction records, resulting in increased efficiency and safety while maintaining high integrity, reliability, and security. The proposed solution focused on the use of smart contracts to govern and control all interactions and transactions between all stakeholders in the supply chain ecosystem. All transactions were recorded and stored in the blockchain's immutable ledger with links to a decentralized file system (IPFS), thus providing to all a high level of transparency and traceability into the supply chain ecosystem in a secure, trusted, reliable, and efficient manner [18].

Current academic and industrial frontiers on blockchain application in the supply chain, logistics, and transportation management were presented by M. Pournader et al. The authors conducted a systematic review of the literature and discover four major clusters in the co-citation analysis: technology, trust, trade, and traceability/transparency. They used an inductive method of reasoning to discuss the emerging themes and applications of blockchains for supply chains, logistics, and transportation for each cluster, based on the pool of articles included in it. Finally, they went over the main topics for future blockchain research and its applications in industry and services [19].

H. Treiblmaier et al. presented a framework based on four established economic theories, namely principal-agent theory (PAT), transaction cost analysis (TCA), the resource-based view (RBV), and network theory. The authors discussed closing the current research gap on the potential implications of the blockchain for supply chain management (SCM) (NT). These theories can be used to generate research questions that were both theory-based and industry-relevant [20].

Khanfar et al. conducted a systematic review to find the applications of blockchain technology in the manufacturing industries and its associated supply chain. They touched on various parameters for evaluation of social and economic performance. They discussed the benefits of employing blockchain technology in the sustainable process of manufacturing. They also proposed a model that illustrates how blockchain technology can contribute to sustainable manufacturing [21].

M.P. Kramer et al. studied the distinctions between traditional and blockchain supported food supply chain networks in the domain of agri-food. They identified the major coordination mechanisms that are supported differently depending on the selected blockchain platform types, based on an extensive literature exploration. The findings were then compared to agri-food industry use cases. The authors claimed that the properties of supply chain coordination mechanisms vary based on the blockchain platform chosen, and that the chosen platform has an impact on the economic performance of the proposed business model [22].

J. Nurgazina et al. aimed to highlight the existing practical uses of distributed ledger technology and internet of things in food supply chain. The authors also mentioned the problems of implementation and relevant research topics for future research, to contribute to the United Nations Sustainable Development Goals. The content of over sixty scholarly articles was reviewed in a comprehensive literature analysis, outlining the strategies to solve the difficulties of scalability, security, and privacy of distributed ledger technology. The main issues in its implementation were increased cost, absence of regulatory framework, and the energy consumption of distributed ledger technology [23].

V. Varriale et al. investigated the effects of employing emerging technologies in improving the sustainable process. Three parameters of supply chain process were compared in different virtual environments. In the first case, the conventional process was used without emerging technologies. In the second case, the same process was carried out with blockchain and internet of things. The findings revealed that performance was better when emerging technologies were employed [24].

A. Park et al. checked the long-term reliability of supply chain management (SCM) in terms of environmental and social factors. They evaluated up to which percentage blockchain technology can better reliability parameters for supply chains. They compiled the papers by analyzing with respect to different components of the literature. They discovered that research on the influence of distributed ledger technology on reliability has been expanding regularly [25].

The pandemic situation around the world exposed the hidden vulnerabilities of the supply chain ecosystem. The primary concern of researchers around the world was to increase the capability of the supply chain process. W. Yin et al. went through an extensive study of the literature to find all the possibilities that could improve the capability of the supply chain process. They analyzed the use of blockchain technology in the supply chain and found that blockchain technology had a great potential that could impact the capability of supply chain in multidimensional domains [26].

V. Varriale et al. explored the use case of distributed ledger technology for a reliable supply chain ecosystem. They synthesized the current state of development and proposed a methodology for the future development. Implications of the research were also investigated. They also explained the advantages of integrating blockchain technology for the development of a reliable supply chain [27].

After the study and analysis of the supply chain ecosystem related literature survey, we found that the domain of the supply chain supported by emerging technologies has been less explored as compared to other domains. We also found that research papers are mostly theoretical and very little information is provided at the implementation level. Some of the authors did try to explain the implementation point of view. Therefore, this study was designed to explore the use of blockchain technology in the supply chain ecosystem. We also investigated the implementation perspective. We used the Ethereum blockchain with proof of authority consensus algorithm, the smart contract written in solidity, and the interplanetary file system for decentralized storage. As the digital environment and technology are evolving, the supply chain ecosystem should be updated with the new emerging technologies.

3. Blockchain Technology for the Supply Chain

Blockchain technology is one of the emerging technologies that will affect most of the domains which are working in a collaborative environment. It is a distributed ledger technology in which different entities can communicate with each other in a peer-to-peer network without any middle entity [28]. It provides multiple features (shown in Figure 2) that can bring trust and transparency in a collaborative environment.

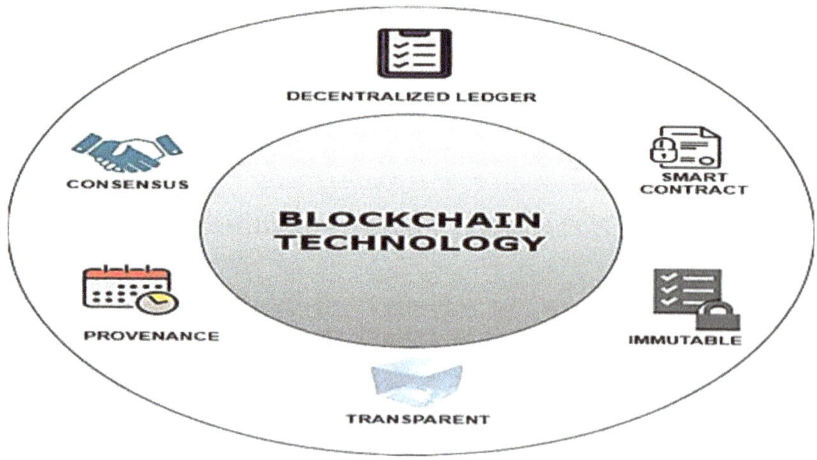

Figure 2. Blockchain Technology features.

Because of its decentralized nature, no single entity can control it. Consensus is the agreement of all the stakeholders on the execution of a transaction. It provides a provenance feature by which the history of any transaction can be traced back. All the transactions are updated in the distributed ledger so that these are visible to all the connected nodes of the network [29]. One of the most important features that make it tamper-proof is immutability. Once any transaction is written in the distributed ledger then no one can edit or delete this transaction. It also supports smart contracts. A smart contract is a computer program that contains the logic of a contract between two or more entities. This contract is executed when certain conditions are met [30].

Advantages of Integrating Blockchain Technology in the Supply Chain

- **Transparent procurement**: Whenever any company goes for procurement, it looks for some middle entity that can do procurement for them. In the traditional approach, it is very difficult to track the actual volume or quantity among all the stakeholders such as partner firms, subsidiaries, etc. The blockchain's distributed ledger can help in this process of procurement. This ledger is shared among all the stakeholders and continuously updated so that each transaction is visible to all. Companies can easily verify their orders by a distributed ledger. Companies have to recruit many people for auditing purposes only. With the help of blockchain, auditing can be done within minutes without the involvement of so many people [31].
- **Smart contracts for payments**: Normally whenever a task is completed an invoice is generated and sent to the client for the payment of the task. As this process is operated manually it can take more time than the expectation of the concerned party. By using blockchain, smart contracts can be integrated into this process. Whenever a task is completed, its invoice is generated automatically and sent to the client digitally. After the verification of the invoice, the smart contract will be triggered, and payment is automatically credited into the account of the company. This is how the integration of blockchain can improve the process of payment settlement [32].
- **No more fraud by rogues**: With blockchain, a decentralized ledger is shared among all the connected nodes. All the transactions are updated in this ledger which is reflected in the ledgers of every connected node. If a rogue tries to execute some transaction for his benefit, he will not succeed. Because blockchain employs a consensus feature, no transaction can be executed without the agreement of all the stakeholders [33].
- **Provenance tracking**: If there is any case where historical data of any transaction is needed to resolve any discrepancy then blockchain technology can be useful. The blockchain-supported solution has a feature of provenance tracking. It means any transaction can be traced back to its time of origin from its current instance [34].
- **Immutable transactions**: There can be multiple transactions executed by authorized nodes. If any attacker node wants to delete or modify the transactions, he would not be able to do this. Because blockchain has an immutability feature once a transaction is written in the decentralized ledger nobody can delete or modify the transaction. Even the administrator cannot delete or update the previously executed transactions. As hash function is used in blockchain technology, even a small change in the data can change the hash of the data as shown below. If only H is replaced by h, then the hash is completely changed [35].

Hash (Hello) = 185f8db32271fe25f561a6fc938b2e264306ec304eda518007d1764826381969
Hash (hello) = 2cf24dba5fb0a30e26e83b2ac5b9e29e1b161e5c1fa7425e73043362938b9824

4. Methodology and Proposed Model

We have proposed a model that uses blockchain technology and associated concepts for the betterment of the supply chain ecosystem. In the traditional system, the supply chain ecosystem begins with the producer of the product and ends with the consumer of the product. A producer produces a product and sells it to a processor for processing of data at the next level of the supply chain ecosystem. Then that processor processes the raw product and converts it to a usable form. Then the processor sells it to a distributor at the next level of the supply chain system. Then the distributor sells it to retailers at the next level. Finally, the retailer sells the product to the consumer of the product. In this traditional approach, a centralized approach is used. Only communicating parties at each level share the data and stakeholders at other levels are not able to see the transactions. In this traditional approach, the history of the product is not traceable so the end user cannot verify the source of the product. Also, the data are not trustworthy because this approach is centralized. In our proposed model, blockchain technology is employed which is decentralized in nature. Decentralized means no single entity will be able to control the complete supply chain ecosystem. There will be no client server approach as it supports a peer-to-peer network model. Any entity can communicate with any other entity on the network without involving any intermediaries. All the data will be stored in a decentralized manner. If some of the systems are down, then it will not affect the performance of the network. All the transactions are visible to all stakeholders. Historical data is also retained in the blockchain so the consumer can easily verify the source of the product. In our proposed model, blockchain is utilized to provide security to documents such as purchase orders, invoices, etc. uploaded by the respective stakeholder of the supply chain by using Ethereum blockchain and IPFS.

The blockchain supported supply chain ecosystem is shown in Figure 3. The main components of the proposed model are digital documents, Ethereum blockchain, interplanetary file system, and system users. All the components are integrated so that these components can exchange information with each other. These components are discussed in the following.

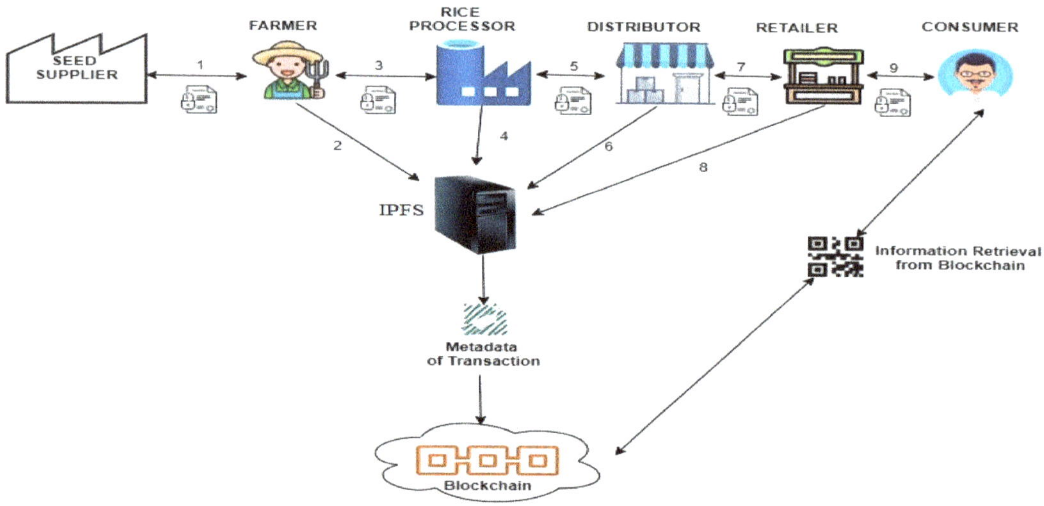

Figure 3. Blockchain Supported Supply Chain Ecosystem.

(1) Digital Documents: These are the documents like purchase orders, dispatch advice, payment advice, etc. that are uploaded by different stakeholders such as the manufacturer, distributor, retailer, etc. during the process of production of goods to their delivery to consumers [36]. This is one of the most important components of the proposed model. Multiple entities will access these documents during the process. so, the security and integrity of these documents are of prime concern because tampered with or manipulated documents can lead to fraud. Such fraud can cause a big loss to stakeholders that will hurt their trust. So, decentralized access control is provided with the help of the Ethereum blockchain which is the next component of the proposed model.

(2) Ethereum Blockchain: Ethereum is an open source blockchain that is used to develop decentralized application. It uses a ledger that stores the transactions. This technology can be used to store the access transactions of documents. Because of the immutability property of the blockchain, document access transactions cannot be modified.

Documents cannot be stored on the Ethereum blockchain because the size of the block in the blockchain is very small. Off-the-chain storage is required to store the digital documents. IPFS, which is discussed next, is used to fulfil our need for off-the-chain storage.

(3) IPFS: Interplanetary File System is a distributed system for storing and accessing information such as documents, websites, applications, etc. It uses a content addressing scheme by which any file can be addressed by its content. Whenever a file is stored on IPFS, a unique cryptographic hash is returned. This hash is used to address or identify that file. This hash also avoids duplication of data because the hash remains the same for two files with the same data. Digital documents will be stored in encrypted form on IPFS in the proposed model [37].

(4) System User: The system user can be any stakeholder of the process such as the consumer, retailer, distributor, or manufacturer. Each user of the system is provided with a pair of keys that can be used while accessing the digital documents related to them.

5. Work-Flow Analysis

We now take the example of supplying rice from farmer to consumer. In this example, first, farmers buy seeds from the seed supplier. Then he grows seeds on his farm. After cutting the crops he will upload the cropped detail (CD) document which contains details like seed type, seed supplier and crops, etc. on the IPFS. Information related to this document will be stored on the blockchain in the form of a transaction. Then rice processor can place a buy order (BO) directly to the farmer after checking all the details of seeds and crops. After receiving the unprocessed rice, the processor processes the rice and uploads the document which contains the processing details. The distributor places a BO to the processor. After receiving the processed rice, the distributor checks for a BO from the retailer. Then the distributor ships the order to various retailers. The distributor uploads a document that contains shipment details of the order. After receiving the rice, the retailer sells the rice to the consumer. The consumer can check the complete data related to the rice from its beginning date to end date. Each entity can communicate with another entity with the help of a smart contract. This model will use the proof of authority consensus mechanism. According to this consensus, only selected nodes are given the responsibility of validating the transactions. For example, if any entity wants to upload data, then the request will be sent to the validator node to validate the transaction. The validator node will receive a notification regarding the transaction. If the validator finds it to be a legitimate transaction, then he approves the transaction. If the validator finds it to be non-authentic then he can decline the transaction. The entity-relationship diagram is shown in Figure 4.

The sequence diagram for the proposed model is shown in Figure 5. This diagram shows the sequence of function execution during the supply chain process of rice from its origin to its delivery to the consumer.

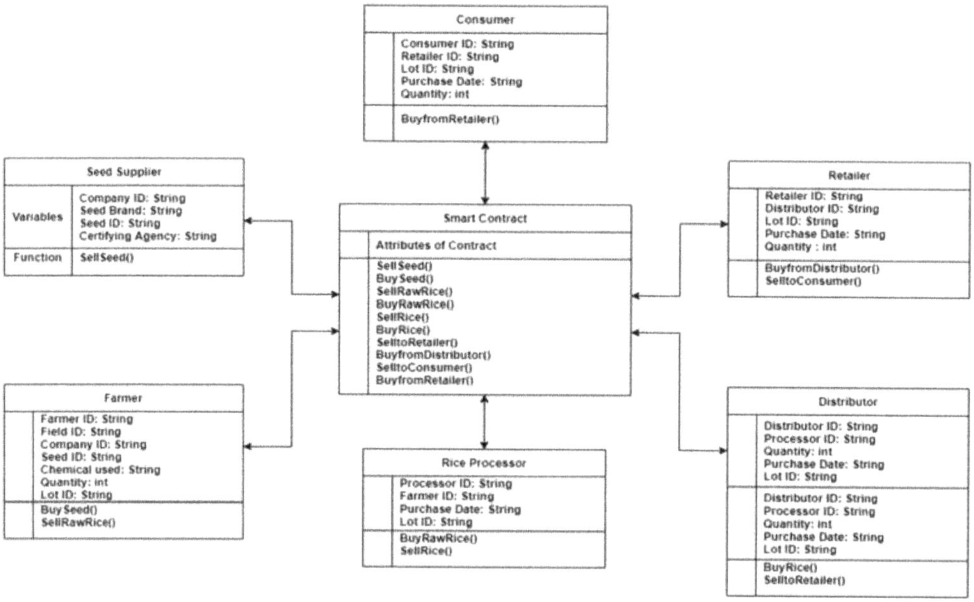

Figure 4. Entity Relationship Diagram for the proposed model.

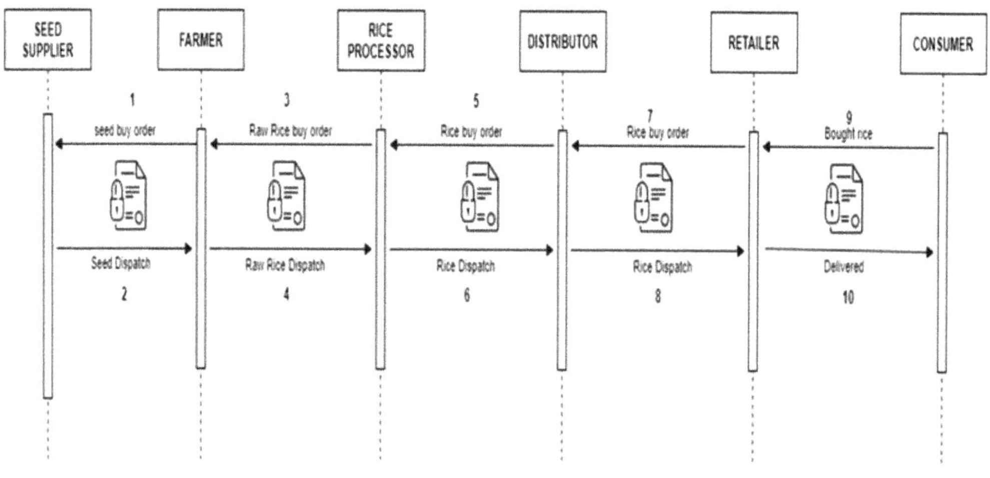

Figure 5. Sequence Diagram for the proposed model.

Two algorithms for communication between seed supplier, farmer, and rice processor are given below. Algorithm 1 illustrates the smart contract communication between the seed supplier and the farmer. Seed is requested from the seed supplier. If the seed sale is agreed and seed prices are paid then the seed request is granted and a notification is sent to the farmer. Otherwise, the request is rejected, and notification of the cancelled request is sent to the farmer. In Algorithm 2, smart contract communication between the farmer and the rice processor is shown. In this algorithm a similar process is followed. If the raw rice sale is agreed and the price is paid then request of the rice processor is granted or else rejected.

Algorithm 1. Farmer Buys Seeds from the Seed Supplier

The ID of Registered Seed Supplier,
Seed Brand,
Seed ID
Certifying agency
Farmer ID, Field ID, Chemical used, Lot ID, Quantity, purchase date
1 Contract state is buy Seed From Seed Supplier
2 State of the Farmer is Seed Requested
3 Seed Supplier state is Wait for Sell Seed To Farmer
4. Restrict access to only registered Seed Supplier
5 if Seed Sale is agreed and Seed Price paid then
6 Contract state changes to Seed Requested Agreed
7 Change State of the Farmer to
Wait For Seed from Seed Supplier
8 Seed Supplier state is Sell Seed to Farmer
9 Send a notification of seed sale to the farmer
10 end
11 else
12 Contract state changes to Seed Request Failed
13 State of a farmer is Request Failed
14 Seed Supplier state is Cancel Request of Farmer
15 Send a notificatfion stating request failure
16 end
17 else
18 Reset contract and displays an error message.
19 end

Algorithm 2. Farmer Sell Raw Rice to Rice Processor

ID of registered Rice Processors
Farmer ID,
Seed ID,
Field ID,
Quantity,
Purchase Date,
Raw Rice Price
1 Contract state is Sell Raw Rice to Rice Processor
2 State of the Rice Processor is Raw Rice Requested
3 Farmer state is Wait for Sell Raw Rice To Rice Processor
4 Restrict access to only Registered Processor
5 if Raw Rice Sale is agreed and Raw Rice Price paid then
6 Contract state changes to Raw Rice Request Agreed
7 Change State of the Rice Processor to
Wait For Raw Rice from Farmer
8 Farmer state is Sell Raw Rice to Rice Processor
9 Send a notification of Raw Rice sale to Rice Processor
10 end
11 else
12 Contract state changes to Raw Rice Request Failed
13 State of Rice Processor is Request Failed
14 Farmer state is Cancel Request of Rice Processor
15 Send a notification stating request failure
16 end
17 else
18 Reset contract and displays an error message.
19 end

The plus of this approach over the standard food supply chain systems is shown in Table 1.

Table 1. Outcomes of the proposed model over traditional models.

Attributes	Proposed System (with Blockchain)	Traditional Method (without Blockchain)
Structure	Decentralized	Centralized
Storage	Distributed	Centralized
Provenance	History of transactions are available	Historical information is not recorded
Immutability	Once written, even admin cannot change the transaction information	The administrator can easily manipulate the information
Trust	Trust in a collaborative environment is increased	Trust issues are there with a centralized approach
Consensus	Agreement between stakeholders is considered.	No such feature is available

The proposed model takes full advantage of the blockchain's benefits, overcomes the problem of depending on core companies to gather information, and makes the information interaction between all nodes more transparent. The issue of unauthorized tampering is avoided because the blockchain information cannot be tampered with. Furthermore, utilizing the blockchain's consensus process, the 'trust problem" between different links in the conventional food supply chain is overcome.

In previous work, some authors did not give the name of the consensus algorithm and some authors used the byzantine fault tolerant algorithm such as proof of work. The proof of work algorithm is time consuming as it takes approximately 10–60 min to add a transaction to a blockchain. It needs to solve a mathematical puzzle to make a transaction so resource consumption is high. It has a risk of 51% of attack. With 51% attack, if an entity is able to control 51% or more nodes of the blockchain network then the proof of work consensus algorithm will be of no use. In our proposed model, the proof of authority consensus algorithm is used. In this algorithm, a few selected nodes can act as validator to validate the transactions. Validators are nodes that have the same identity on the network as they do in the public notary database. They reveal their true identities willingly. Validators demonstrate their commitment to the network over time. They should be willing to invest their money and risk their reputation to maintain the network running smoothly. In our model, representatives from the seed supplier or rice processor company or a distributor can act as validator nodes. All the transactions will be validated by these validator nodes. The time required to generate a block is predictable and less than the proof of work algorithm because rather than involving all the nodes in the validation of transactions, only preselected validator nodes will verify the transaction. This model encourages collaboration and verification among grain supply chain stakeholders, integrates resources efficiently, and optimizes benefits. After a thorough study of the literature, a comparative analysis of traditional models with the proposed model is shown in Figure 6. As per our observation, we found that the proposed system is far better than the traditional method in all attributes. These scores represent the presence or absence of features such as trust, immutability, provenance, and consensus. A high score represents a significant presence of the particular feature in the proposed model and a low score for the traditional models represents the absence or nominal presence of the particular feature. No experts were used for the evaluation. Comparative analysis compares traditional models and the proposed model and determines the presence and absence of the above-mentioned features. Comparison is done on the basis of various attributes such as trust, immutability, provenance, and consensus. The proposed model is more trustworthy as it works in a decentralized manner because of blockchain technology. There is no central authority that can control the entire network but the model uses immutable distributed ledger therefore once the transaction is written in the ledger it cannot be deleted or modified. This model has a provenance feature. It means transactions can be traced back from their current state to their origin state and also has a consensus mechanism whereby every transaction will be executed by the agreement of the authorized nodes. High scores represent a significant presence of the respective parameter and low scores represent the absence of the same.

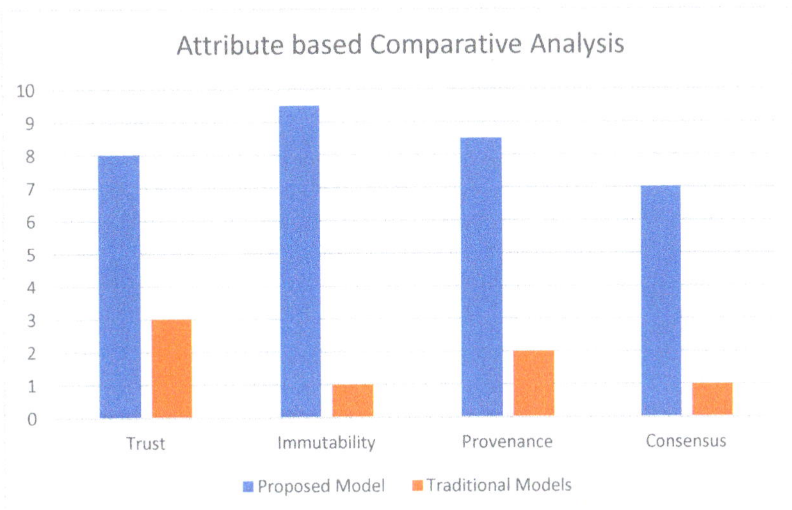

Figure 6. Attribute based comparative analysis.

The limitations of the blockchain technology-based model are that it has constraints such as the number of transactions per second, latency in transactions, and the amount of data in transactions. Regular updates in blockchain technology frameworks will surely improve the performance of these types of models in the future.

6. Conclusions

In a multi-stakeholder supply chain, maintaining product and process integrity is a significant challenge. Data fragmentation, a lack of reliable provenance, and a variety of protocol regulations across multiple distributions and processes plague many current solutions. Blockchain, among other technologies, has emerged as a leading technology because it provides secure traceability and control, immutability, and stakeholder trust in a low-cost IT solution [38]. The main goal of this article was to answer two research questions. First, what are the most important issues of the traditional supply chain ecosystem and second, how can these issues be resolved by integrating blockchain technology with the supply chain ecosystem. To determine the issues, we explored multiple research articles of this domain. To answer the second question, the various benefits of integration blockchain technology with the supply chain were explored and a model proposed. The proposed model employs the Ethereum blockchain with the proof of authority consensus algorithm and an interplanetary file system. This proposed model improves the performance by removing intermediaries and making payment processing via smart contract. We presented different aspects of the proposed model such as entity-relationship diagrams, sequence diagrams, and algorithms for smart contract communication. We presented how this integration of blockchain technology in the supply chain can bring trust, transparency, and security by immutability. This model can be applied to different supply chain models of various domains with slight amendments.

Author Contributions: Conceptualization, S.K.R. (Sumit Kumar Rana), A.K.R., S.K.P., S.K.R. (Sanjeev Kumar Rana); methodology, S.K.R. (Sumit Kumar Rana), A.K.R., S.K.P., S.K.R. (Sanjeev Kumar Rana); validation, S.K.R. (Sumit Kumar Rana), A.K.R., S.K.P., S.K.R. (Sanjeev Kumar Rana); formal analysis, S.K.R. (Sumit Kumar Rana), A.K.R., S.K.P., S.K.R. (Sanjeev Kumar Rana); data curation, S.K.R. (Sumit Kumar Rana), A.K.R., S.K.P., S.K.R. (Sanjeev Kumar Rana); writing—original draft preparation, S.K.R. (Sumit Kumar Rana), A.K.R., S.K.P., S.K.R. (Sanjeev Kumar Rana); writing—review and editing, S.K.R. (Sumit Kumar Rana), A.K.R., and S.A.; supervision, S.A. and H.-C.K.; project administration, S.A., M.-I.J., and H.-C.K.; funding acquisition, S.A., M.-I.J., and H.-C.K. All authors have read and agreed to the published version of the manuscript.

Funding: This research was supported by This research was supported by the Basic Science Research Program through the National Research Foundation of Korea (NRF) funded by the Ministry of Education (NRF2021R1I1A1A01050306) and by the Ministry of Science, ICT, and Future Planning (NRF2017R1D1A3B04032905) and This work was supported by the Commercializations Promotion Agency for R&D Outcomes (COMPA) grant funded by the Korean Government (Ministry of Science and ICT)" (R&D project No.1711139492).

Conflicts of Interest: The authors declare no conflict of interest.

References

1. Omar, I.A.; Jayaraman, R.; Debe, M.S.; Omar, M. Automating Procurement Contracts in the Healthcare Supply Chain Using Blockchain Smart Contracts. *IEEE Access* **2021**, *9*, 37397–37409. [CrossRef]
2. Kumar, A.; Sharma, S.; Goyal, N.; Singh, A.; Cheng, X.; Singh, P. Secure and energy-efficient smart building architecture with emerging technology IoT. *Comput. Commun.* **2021**, *176*, 207–217. [CrossRef]
3. Chang, S.E.; Chen, Y. When Blockchain Meets Supply Chain: A Systematic Literature Review on Current Development and Potential Applications. *IEEE Access* **2020**, *8*, 62478–62494. [CrossRef]
4. Wang, L.; Xu, L.; Zheng, Z.; Liu, S.; Li, X.; Cao, L.; Li, J.; Sun, C. Smart Contract-Based Agricultural Food Supply Chain Traceability. *IEEE Access* **2021**, *9*, 9296–9307. [CrossRef]
5. Rana, A.K.; Sharma, S. Enhanced energy-efficient heterogeneous routing protocols in WSNs for IoT application. *Int. J. Eng. Adv. Technol.* **2019**, *9*, 4418–44250.
6. Lezoche, M.; Hernandez, J.E.; Diaz, M.; Panetto, H.; Kacprzyk, J. Agri-food 4.0: A survey of the supply chains and technologies for the future agriculture. *Comput. Ind.* **2020**, *117*, 103187. [CrossRef]
7. Gopi, K.; Mazumder, D.; Sammut, J.; Saintilan, N. Determining the provenance and authenticity of seafood: A review of current methodologies. *Trends Food Sci. Technol.* **2019**, *91*, 294–304. [CrossRef]
8. Astill, J.; Dara, R.; Campbell, M.; Farber, J.; Fraser, E.; Sharif, S.; Yada, R. Transparency in food supply chains: A review of enabling technology solutions. *Trends Food Sci. Technol.* **2019**, *91*, 240–247. [CrossRef]
9. Zhao, Z.; Dong, Y. Blockchain-Based Safety Management System for the Grain Supply Chain. *IEEE Access* **2020**, *8*, 36398–36410.
10. Della Valle, F.; Oliver, M. Blockchain Enablers for Supply Chains: How to Boost Implementation in Industry. *IEEE Access* **2020**, *8*, 209699–209716. [CrossRef]
11. Shahid, A.; Almogren, A.; Javaid, N. Blockchain-Based Agri-Food Supply Chain: A Complete Solution. *IEEE Access* **2020**, *8*, 69230–69243. [CrossRef]
12. Musamih, A.; Salah, K.; Member, S.; Jayaraman, R. A Blockchain-Based Approach for Drug Traceability in Healthcare Supply Chain. *IEEE Access* **2021**, *9*, 9728–9743. [CrossRef]
13. Kamilaris, A.; Fonts, A.; Boldύ, F.X.P. The rise of blockchain technology in agriculture and food supply chains. *Trends Food Sci. Technol.* **2019**, *91*, 640–652. [CrossRef]
14. Blossey, G.; Eisenhardt, J.; Hahn, G. Blockchain Technology in Supply Chain Management: An Application Perspective. In Proceedings of the 52nd Hawaii International Conference on System Sciences, Grand Wailea, HI, USA, 8–11 January 2019; Volume 6, pp. 6885–6893.
15. Saberi, S.; Kouhizadeh, M.; Sarkis, J.; Shen, L. Blockchain technology and its relationships to sustainable supply chain management. *Int. J. Prod. Res.* **2018**, *57*, 2117–2135. [CrossRef]
16. Gonczol, P.; Katsikouli, P.; Herskind, L.; Dragoni, N. Blockchain Implementations and Use Cases for Supply Chains-A Survey. *IEEE Access* **2020**, *8*, 11856–11871. [CrossRef]
17. Tsang, Y.P.; Choy, K.L.; Wu, C.H.; Ho, G.T.S.; Lam, H. Blockchain-Driven IoT for Food Traceability With an Integrated Consensus Mechanism. *IEEE Access* **2019**, *7*, 129000–129017. [CrossRef]
18. Salah, K.; Nizamuddin, N.; Jayaraman, R.; Omar, M. Blockchain-Based Soybean Traceability in Agricultural Supply Chain. *IEEE Access* **2019**, *7*, 73295–73305. [CrossRef]
19. Pournader, M.; Shi, Y.; Seuring, S.; Koh, S.C.L. Blockchain applications in supply chains, transport and logistics: A systematic review of the literature. *Int. J. Prod. Res.* **2020**, *58*, 2063–2081. [CrossRef]
20. Treiblmaier, H. The impact of the blockchain on the supply chain: A theory-based research framework and a call for action. *Supply Chain Manag. Int. J.* **2018**, *23*, 545–559. [CrossRef]

21. Khanfar, A.A.A.; Iranmanesh, M.; Ghobakhloo, M.; Senali, M.; Fathi, M. Applications of Blockchain Technology in Sustainable Manufacturing and Supply Chain Management: A Systematic Review. *Sustainability* **2021**, *13*, 7870. [CrossRef]
22. Kramer, M.P.; Bitsch, L.; Hanf, J. Blockchain and Its Impacts on Agri-Food Supply Chain Network Management. *Sustainability* **2021**, *13*, 2168. [CrossRef]
23. Nurgazina, J.; Pakdeetrakulwong, U.; Moser, T.; Reiner, G. Distributed Ledger Technology Applications in Food Supply Chains: A Review of Challenges and Future Research Directions. *Sustainability* **2021**, *13*, 4206. [CrossRef]
24. Varriale, V.; Cammarano, A.; Michelino, F.; Caputo, M. Sustainable Supply Chains with Blockchain, IoT and RFID: A Simulation on Order Management. *Sustainability* **2021**, *13*, 6372. [CrossRef]
25. Park, A.; Li, H. The Effect of Blockchain Technology on Supply Chain Sustainability Performances. *Sustainability* **2021**, *13*, 1726. [CrossRef]
26. Yin, W.; Ran, W. Theoretical Exploration of Supply Chain Viability Utilizing Blockchain Technology. *Sustainability* **2021**, *13*, 8231. [CrossRef]
27. Varriale, V.; Cammarano, A.; Michelino, F.; Caputo, M. The unknown potential of blockchain for sustainable supply chains. *Sustainability* **2020**, *12*, 9400. [CrossRef]
28. Rana, A.K.; Sharma, S. Industry 4.0 manufacturing based on IoT, cloud computing, and big data: Manufacturing purpose scenario. In *Advances in Communication and Computational Technology*; Springer: Singapore, 2021; pp. 1109–1119.
29. Rana, A.K.; Sharma, S. Contiki Cooja Security Solution (CCSS) with IPv6 Routing Protocol for Low-Power and Lossy Networks (RPL) in Internet of Things Applications. In *Mobile Radio Communications and 5G Networks*; Springer: Singapore, 2021; pp. 251–259.
30. Buterin, V. A next generation smart contract & decentralized application platform- Ethereum white paper. *White Paper* **2013**, *3*, 1–36.
31. Khan, S.; Loukil, F.; Guegan, C.G.; Benkhelifa, E.; Bani hani, A. Blockchain smart contracts: Applications, challenges and future trends. In *Peer-to-Peer Networking and Applications*; Springer: Canada, 2021. Available online: https://www.ncbi.nlm.nih.gov/pmc/articles/PMC8053233/ (accessed on 30 August 2021).
32. Norvill, R.; Pontiveros, B.B.; State, R.; Cullen, A. IPFS for reduction of chain size in Ethereum. In Proceedings of the IEEE Conference on IoT, Green Computing, Smart Data, Blockchain, Congress on Cybernetics, Halifax, NS, Canada, 30 July–3 August 2018; pp. 1121–1128.
33. Dias, D.; Benet, J. *Distributed Web Applications with IPFS in ICWE*; Springer LNCS: Basel, Switzerland, 2016; pp. 616–619.
34. Kouhizadeh, M.; Saberi, S.; Sarkis, J. Blockchain technology and the sustainable supply chain: Theoretically exploring adoption barriers. *Int. J. Prod. Econ.* **2020**, *231*, 107831. [CrossRef]
35. Nandi, S.; Sarkis, J.; Hervani, A.A.; Helms, M.M. Redesigning Supply Chains using Blockchain-Enabled Circular Economy and COVID-19 Experiences. *Sustain. Prod. Consum.* **2020**, *27*, 10–22. [CrossRef]
36. Kurpjuweit, S.; Schmidt, C.G.; Klöckner, M.; Wagner, S.M. Blockchain in Additive Manufacturing and its Impact on Supply Chains. *J. Bus. Logist.* **2019**, *42*, 46–70. [CrossRef]
37. Saurabh, S.; Dey, K. Blockchain technology adoption, architecture, and sustainable agri-food supply chains. *J. Clean. Prod.* **2020**, *284*, 124731. [CrossRef]
38. Swain, S.; Peter, O.; Adimuthu, R.; Muduli, K. Blockchain technology for limiting the impact of pandemic: Challenges and prospects. In *Computational Modeling and Data Analysis in COVID-19 Research*; CRC Press: Boca Raton, FL, USA, 2021; pp. 165–186.

sustainability

MDPI

Article

E-Agricultural Supply Chain Management Coupled with Blockchain Effect and Cooperative Strategies

Mohammed Alkahtani [1,2,*], Qazi Salman Khalid [3], Muhammad Jalees [4], Muhammad Omair [5], Ghulam Hussain [6] and Catalin Iulian Pruncu [7,8,*]

1 Industrial Engineering Department, College of Engineering, King Saud University, Riyadh 11421, Saudi Arabia
2 Raytheon Chair for Systems Engineering (RCSE), Advanced Manufacturing Institute, King Saud University, Riyadh 11421, Saudi Arabia
3 Department of Industrial Engineering, University of Engineering and Technology, Peshawar 25000, Pakistan; qazisalman@uetpeshawar.edu.pk
4 Department of Management Sciences, Bahria University, E-8, Islamabad 44000, Pakistan; muhammadjalees123@gmail.com
5 Department of Industrial Engineering, Jalozai Campus, University of Engineering and Technology, Peshawar 25000, Pakistan; muhamad.omair87@gmail.com
6 Faculty of Mechanical Engineering, GIK Institute of Engineering Sciences & Technology, Topi 23460, Pakistan; gh_ghumam@hotmail.com
7 Department of Mechanical Engineering, Imperial College London, Exhibition Rd., London SW7 1AY, UK
8 Design, Manufacturing & Engineering Management, University of Strathclyde, Glasgow G1 1XJ, UK
* Correspondence: moalkahtani@ksu.edu.sa (M.A.); c.pruncu@imperial.ac.uk (C.I.P.)

Abstract: The agricultural industry is highly underdeveloped and requires transformation in technology for food safety and reliability. A digital world is relying on blockchain technology for the successful implementation of sustainable e-agricultural supply chain management (e-Agri-SCM). In current advancements of blockchain in digital marketing, product website design (web design) is essential to streamline the requirements of the customer and the expectations of supply chain partners. The current research has incorporated the blockchain effect by web design elements into the agricultural supply chain management (Agri-SCM) study. In addition, partners in the digital marketing supply chain (DM-SCM) are also facing issues to identify significant web design elements-based blockchain technology to gain maximum profit. Therefore, a cooperative (Co-op) sustainable e-agricultural SCM model is developed in this study by considering the web design index and variable demand to decide shipments, selling price, cycle time, and advertisement cost for agriculture products. The uncertainties in the model due to intangible web design elements and basic costs are dealt with by the application of the fuzzy system whereas carbon emission is also considered for providing cleaner production. A real-time application of the proposed model is done by undertaking five different cases based on mutual share, demand curve, and advertisement budget among participants. The sensitivity analysis is also performed to identify important factors of the total profit. Findings of this work include significant web design elements (WDEs) i.e., web graphics, search engine optimization, cyber-security, fast loading, and navigation, as essentials for digital marketing to convince customers towards the product in a global SCM. The numerical results and managerial insights are advantageous for managers to get maximum profit by cooperative and digital marketing strategies to attain e-Agri-SCM.

Keywords: sustainable agricultural supply chain management; web design elements; blockchain; variable demand; cooperative advertisement; uncertain environment

Citation: Alkahtani, M.; Khalid, Q.S.; Jalees, M.; Omair, M.; Hussain, G.; Pruncu, C.I. E-Agricultural Supply Chain Management Coupled with Blockchain Effect and Cooperative Strategies. *Sustainability* **2021**, *13*, 816. https://doi.org/10.3390/su13020816

Received: 8 December 2020
Accepted: 10 January 2021
Published: 15 January 2021

Publisher's Note: MDPI stays neutral with regard to jurisdictional claims in published maps and institutional affiliations.

Copyright: © 2021 by the authors. Licensee MDPI, Basel, Switzerland. This article is an open access article distributed under the terms and conditions of the Creative Commons Attribution (CC BY) license (https://creativecommons.org/licenses/by/4.0/).

1. Introduction

Owing to the escalating awareness of resource depletion, climate change, and overwhelmed population, firms in the agriculture domain need to redesign their current supply chain models to be sustainable by taking economic and environmental impacts into account. Drastic evolution has been observed in the agriculture supply chain during the last four decades [1]. Production technology and marketing has played a vital role in the upgradation agriculture supply chain and has been a limelight for governments and agri-business sectors. Further, the development in information, its transmission, and transportation mediums have evolved. Currently, they are encapsulated with vigorous internet usage through various platforms worldwide [2]. Agri business, specifically attached with modern digitized technologies such as blockchain have revolutionized the medium of business conducted with customers world-wide, by providing social connections that vitally help in searching, targeting, communicating, and serving buyers [3]. Firms may observe tremendous benefit by enhancing investments in technology, such as higher efficiency and lower costs, and high service level [4]. Moreover, the boom in technology has strengthened the online shopping sector. According to a report published by the National Bureau of Statistics (2016), the gross online merchandise volume of China amounted to 4.8 trillion Yuan, increasing by 64.5 percent in just a year span. This growth was recorded to be four times more rapid than cumulative retail sales of goods [5].

Digitization and online marketing are strongly corelated with blockchain technology. Blockchain technology emanated by the development of digital currency—Bitcoin that appeared initially in the form of Bitcoin white paper [6]. The benefits of blockchain have encouraged the business communities and industries to adopt it, which is reflected from a survey that exhibits 77% respondents (proficient in blockchain) from a set of 600 executives acknowledged their firm's involvement in the blockchain projects [7]. Blockchain technology permits the construction of an encrypted and secured set of records between parties digitally connected in a supply chain [8]. Blockchain consist of transactions or blocks of information stored in archives available on a mutual platform to the supplier and vendors. These data blocks require nodes which are services connected through the internet and websites.

Furthermore, investment in the right direction for technologies is very critical. For this, Busca and Bertrandias [9] developed a framework for digital marketing (DM) and investigated its four cultural eras, and compared with the effect of blockchain. They devised different plans for the potential evolution of DM and in view of these plans, website or application design is a significant element to attract the customers and prosper e-business. Blockchain integrated with efficiently designed website increases probabilities of two major factors; firstly, gaining attraction of the customers and secondly traceability of the products from initiation to the terminal points. A website can transmit intrinsic product features like a written context of the product, images, and virtual experiences as well as extrinsic product features such as brand and price [10]. From a customer point of view, the element of a website has transformed into a pivot point that assesses the objectives of the system, service, and information [11]. On the contrary, from a product or service providers' point of view, the elements of a website transformed should be a communal distribution medium to generate profit. Empirical studies on websites linking to the digital marketing supply chain (DM-SC) is rare. Owing to the rapid expansion in e-commerce around the globe, the e-store environment has fascinated maximum attraction from business during the past 12 years [12].

To narrow down our research stream, the grooming technology via e-marketing depends on the integration of blockchain in SC, developing website esthetics and the contents significantly. Interestingly, a report was published in a UK-based study, analyzing the inspiration element of trust among customers and retailers on websites, specifically over the impact of web content in comparison to web design. It was staggering to conclude that 94% of customers mistrusted a website directly due to design elements and 6% showed mistrust based on the actual content of the website [13]. Hence, the quality of a website is essential during the current digitized global era. As there was no study found on the integration of website esthetics including the effect of blockchain with DM-SC, the current research targets the need to fill this gap.

The article is structured by introducing the web designing of agri-product and digital marketing encapsulated with blockchain, to provide a platform for a sustainable and robust e-agriculture supply chain management (SCM) in this section. The related research works and reviews are represented in Section 2. The research methodology and framework including analytically hierarchical process (AHP), Pareto analysis, and fuzzy inference system (FIS) are discussed in detail in Section 3. The mathematical model of proposed cooperative e-agriculture SCM is formulated and modeled in Section 4. Section 5 represents numerical experimentation in which the data of supplier, agri-processing firm, and multi-retailer in SCM are incorporated. In addition, solution methodology and results findings by considering five cases of SCM are well summarized in Section 6. The sensitivity analysis in Section 7 is performed to check the effect of the web design index on total profit of the SCM. Lastly, Section 8 concludes the research study on the basis of finding results.

2. Literature Review

In order to fulfil the requirements of the future generations, agri-product supply chains should lay emphasis on digitized marketing through an effective e-advertisement channel. These resources acquire usage of architectural website designs that attract the customers and smooth the flow of information by using blockchain. Furthermore, a digitized sustainable agri-product supply chain needs more than only an economic validation objective (profit), it should also be keen to handle trust of the linked parties. Agriculture trends should shift towards the adoption of DM as it is less costly and risky in operations, even boundary-less, which allows the firm to deal more proficiently in complex, diverse, and distant markets, while at the same time dependence on domestic infrastructural system is reduced [14]. Firms tend to apply digital technologies currently; tools and devices to circumvent barriers such as pertaining to location and foreign market analysis, communication with foreign customers, and overseas opportunities identification, which were viewed as potential hindering subjects in the engagement, operation, and expansion of a company in international markets [15].

In the recent decade, blockchain technology has been on the rise, gaining success and proving its functionality in various organizations. It aims to harness its limpidity and transparency in order to provide solution to existing problems, where many untrusted actors are indulged in the resource distribution [16]. Agriculture supply chain is highly relevant to the resource distribution since the items of agriculture are predominantly used as an input in several multi-actor supply chains, where consumer is generally the end client. Further, in SCM, the annual growth of blockchain is anticipated with a rate of 87%, an increase from $45 million (2018) to $3314.6 million (2023) [17].

As an effective example, AgriDigital company in December 2016, implemented a pioneer settlement of the sale with 23.46 tons of grain via blockchain. After that, over 1300 different users with supplementary 1.6 million tons of grain got transacted through a cloud system, including a figure of $360 million in the growers' payment. This success provided an inspiration for a potential and substantial usage of this technology in an agri-supply chain. This realm of digitization created enormous room for an e-market platform, which is essential to grab in a competitive environment. E-marketing is an emerging concept that depicts the process of buying and selling or exchanging either/both products or services through the internet [18]. According to Krishnamurthy and Singh [19], the services of e-marketing include a number of customer-support-activities; including e-tailing (direct-selling), SCM, and CRM. E-market services are interactive, dynamic, and internet-based that are sandwiched with customer care applications [20,21]. Websites and social application platforms act as a liaison and an interactive interface between various levels of the supply chain.

Several studies like Cannon and Perreault Jr [22], Albrecht et al. [23], Kalvenes and Basu [24] have endorsed that business-to-business electronic market place has developed to be an ideal podium for both buyers and sellers. E-marketing has a significant contribution to firms' development. Researchers have identified that managing better electronic customer relationship [25], providing effective operative services [26], being adaptive to external knowledge, up to the mark utilization of e-marketing tools [27], and exploration of state-of-the-art information about available e-market [28] leads to effective firm performance. A multi-objective optimization model based on IoT and block chain in e-market was studied by [29] while data preserving in smart agriculture for an e-market was carried out by [30]. The sustainability perceptions of manufacturing stakeholders and agriculture irrigation resources were discussed by [31,32] respectively. The detailed contribution of the researchers is given in Table 1, where a research gap has been filled through proposed research work. Dadzie et al. [33] applied data reliability tests on the web design elements and determined the most effective ones for a business perspective. Likewise, Harrison and Waite [34] also established effective web design elements by applying multivariate analysis of variance (MANOVA) in their parametric study. Afterwards, Hu and Chen [35,36] conducted similar study by applying analysis of variances (ANOVA) and analytical approach respectively to prove significant web design elements. As decision makers are always interested in defining beneficial economic policies, that is where analytical approaches and mathematical models of strategic interaction amid balanced decision-makers come into play, such as, game theory, Nash equilibrium, and cooperative game theory. Established works with these theories and strategies based on a three-echelon cooperative supply chain can be found in the work of [35,37–43]. However, only the study of Cai et al. [41] is based on game theory for the web design elements and its framework does not cover three-echelon SCM. The proposed work matches with Sarkar et al. [44] for the SCM environment and application of fuzzy inference system (FIS), which is used to imply fuzzification procedure to the uncertain factors in the model. However, this work asserts multiple addition with the involvement of addition in the environment by coupling blockchain effect and web design elements. Further, the approach in the current study is also the combination of sequential quadratic programming (SQP), analytical hierarchy process (AHP), and FIS in order to tackle all the uncertain variables with tangible and intangible effects.

Table 1. Author contribution table.

Author	Three-Echelon SCM [1]	Cooperative SCM [2]	Web Design Elements	E-Advert. Cooperation [3]	Variable Demand	Uncertain Conditions	Blockchain Effect	Solution Algorithm
Dadzie et al. [33]			√					Data reliability tests
Harrison and Waite [34]			√					MANOVA
Hu et al. [35]			√					Analytical approach
Chen [36]			√					ANOVA
Taheri et al. [37]		√		√	√			Game theory
Oberoi et al. [38]		√	√		√			Analytical approach
Zhao et al. [39]		√		√	√	√		Nash Equilibrium
Lin et al. [40]	√						√	Analytical approach
Cai et al. [41]			√					Game theory
Kamilaris et al. [42]		√		√		√		Cooperative game approach
Xiao et al. [43]	√	√				√		Meta heuristic
Song et al. [44]		√	√				√	Sustainable data management approach
Choi [45]		√		√		√		Cooperative game approach
Sarkar et al. [46]	√	√		√		√		FIS
Proposed Work	√	√	√	√	√	√	√	AHP-FIS-SQP

[1] Three-echelon supply chain management (SCM) is the management of flow of products through the three-tier chain of composed of manufacturer, retailer, and consumer. [2] Cooperative supply chain management (Co-SCM) is an incentive-grounded supply chain based on" trust", "vision", and "agent"-based models between the stakeholders at each layer of the chain. [3] E-Advert. Cooperation is an electronic advertisement policy under a cooperative supply chain model.

Studies have been carried out related to supply chain decision in e-markets by employing game theory [47], and Stackelberg–Nash equilibrium [48] in a vendor managed inventory including multi-retailers and using retailers' information respectively. The decision model developed by Esmaeili et al. [49] provided factors of pricing and cooperation among sellers and buyers through competitive advertisement. Despite advertisements, a pricing strategy for any product in a sustainable supply chain is vital. Cai et al. [50] worked on pricing policies with price discount transactions in a dual-channel supply chain. Keeping the pricing policy intact, ordering policy is as important for not losing customers. Cai et al. [51] studied pricing along ordering policies in a supply chain model of the Business-to-Business (B2B) market. Further, there are many discount and pricing model studies i.e., Banerjee [52], Dada and Srikanth [53] and Pal et al. [54], where order quantities, price break, lot sizing, and discount quantities are the decision variables. Other than these factors, supplier selection is also important for a sustainable SCM and similar studies are carried out by [55–57].

Despite the fact of digitization and e-market success, advertisement is a significant element to exercise in a meaningful proportion and medium. In the recent decade, several researchers dealt with cooperative advertising in the manufacturer–retailer channel. This collaboration can also be termed as a financial contract, in which a manufacturer offers to accept either a certain portion or the complete advertising expenditure of a retailer/vendor [58]. In that way, increment in the advertising of a retailer is injected by the manufacturer to stimulate an immediate demand in the chain. However, being a potential part of several manufacturers' advertisement budgets (e.g., sum of $15 billion was financed to similar programs in the USA, 2000), a number of forms appear to settle the contribution rate by the rule of thumb or best guess approach, deprived of detailed analysis on proportions of 50% or 100% [59].

The strategies to build up in the SC model comprise of multiple scenarios and attributes (mostly uncertain), thus, for this purpose, their ranking is vital with multi-criteria decision-based systems and fuzzy approaches. Recently, Bhosale and Kant [60] worked on an integrated fuzzy interference system (FIS) and fuzzy Delphi to rank the solutions and overcome the SC barriers. In another study by Amindoust and Saghafinia [61], evaluation of supplier selections in a textile industry was carried out by FIS and suppliers were ranked in order to effectively meet the objectives of selection. Paul et al. [62] applied FIS to predict the variations in demand based on the given forecast. They revised their supply chain plan in advance to develop a mitigated plan. Another study based on ranking of qualitative assessments in a green SC was carried out by [63] with the help of FIS. To sum up, FIS is an important tool for ranking of attributes linked with an uncertain environment. This study is pioneer in the shape of FIS application for ranking website design elements that are linked by an uncertain demand environment in a SC.

In today's world, the agri-product promotion is effectively going on through digital marketing by enhancing web design to attract the customers. The objective is to provide a platform in the form of technology development policy among agricultural businesses for food security and reliability. Researchers are working on the enhancement of the product web design; however, web design is still not digitally connected with the customers through blockchain technology for the successful implementation of global agri-SCM. This study aims to target the e-market (online), which exclusively affects the firms' performance. The significant web design elements are incorporated in cooperative three-echelon SCM under uncertain conditions. The research is specifically devising an attractive digital marketing policy through effective website design attributes, and linking these attributes with customer demand and advertisement cost to boost the total profit of the SCM partners.

3. Material and Methods

The research methodology was devised in a way to ensure a traceability and security to the food customers through blockchain technology in digital marketing. The agri-product web design was developed by incorporating blockchain elements to create digitization in the agri-SCM. The research was based on a step-by-step methodology, from the selection of significant web design elements (WDEs) based on blockchain and basic attributes for the promotion of the product through digital marketing, then the incorporation of the web design index (WDI) into three-echelon SCM, and finally finding the tangible effect of the web design elements on the total profit of the SCM based on co-op e-advertising policy. The research methodology is well represented by the flow diagram given in Figure 1 and the step-by-step procedure was as follows.

1. The first step was associated with the identification and selection of significant web design elements (based on basic and blockchain technology) for the attraction of customer towards product. These significant elements were further considered to find the web design index. Nine basic design elements and six were identified and selected after taking opinion from digital marketing experts for the promotion of the product. Therefore, a questionnaire was developed and distributed among 120 respondents randomly. The response rate was recorded. The data were analyzed using Cronbach's reliability test and it resulted that the α value was greater than 0.7. The basic web design elements were identified as web graphics (GA), hero images (HI), navigation (NA), fast loading (FL), typography (TY), web contents (WC), product videos (PV), branding (B), and card design (CD). On the other hand, blockchain elements to provide advanced digital technology were selected as search engine optimization (SEO), mobile compatibility (MC), machine learning (ML), email marketing (EM), enabled content management system (ECMS), and cyber security (CS). These selected WDEs are processed separately using an analytically hierarchical process (AHP) for the evaluation and ranking of the basic and blockchain design elements. (See Section 3.1).
2. The second step covered the selection of the significant web design elements for the influence of the customer towards product by using Pareto analysis, which is based on an 80-20 rule. Pareto processes the ranking and corresponding eigen values obtained from the AHP. The significant WDEs were selected and considered for the SCM model, which influenced the response (web design index) by 80% and the remaining 20% elements had less impact. As a result, four basic elements and three blockchain elements were selected, which were considered significant for the calculation of the web design index using fuzzy inference system (FIS). (See Section 3.2).
3. The selected combination of three blockchain (i.e., SEO, CS, and ML) and four basic design elements (i.e., GA, NA, FL, TG) were significant design elements and considered as an input in FIS. The uncertainty exists in measuring these significant design elements, which is the reason the FIS application was utilized from MATLAB software to defuzzify the uncertain inputs by using a centroid method and the web design index (WDI) was calculated by formulation of a triangular fuzzy system.
4. Now the data are qualitative and continuous, therefore it was required to convert them into crunch value to find the overall website design index (WDI) on the bases of significant key elements. Therefore, FIS was used as a tool from MATLAB. The rules were taking weights with respect to the customer requirements and technical requirements. Finally, the WDI was calculated as an output of FIS from the combination of the blockchain elements and basic elements of the web design. (See Section 3.3).
5. The WDI exists in a range from one to five depending on how the website is attracting and influencing customer towards product. The WDI as an output of the FIS was also triangular fuzzy number and was considered low between 1.0–2.8, medium between 2.8–3.2, and high within a range from 3.2–5.0. (See Section 3.2).
6. Ultimately, the obtained WDI was incorporated into the agri-SCM for co-op e-advertising among supplier, agri-processing firm, and multi-retailer to grab and capture the e-market shares. The demand of customers from the e-market is also increasing with retailers due to attraction towards product as a result of the improved WDIs. The demand was incorporated into the three-echelon SCM, where the supplier, agri-processing firm, and multi-retailer cooperate based on advertisement cost by enhancing design elements of the product website for influencing customers. (See Section 4).
7. The proposed SCM model was solved and optimized using sequential quadratic optimization (SQP) nonlinear interactive optimization technique. The optimal solution

was based on cycle time, shipments, e-advertisement cost share by SCM partners, and selling price. (See Section 6).

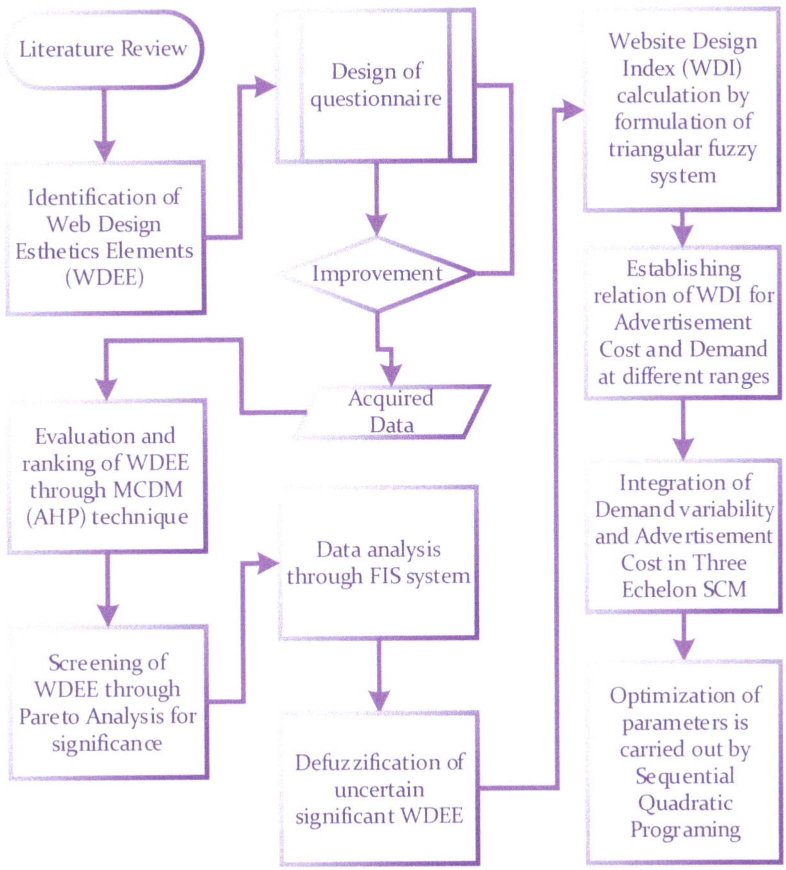

Figure 1. Flow diagram of the research study of digital marketing and blockchain in e-agricultural supply chain management.

3.1. Analytic Hierarchy Process

The nature of the problem is quite different to understand the effectiveness of the product web design to attract customer and improve product demand. The new era is totally dependent on new technologies i.e., blockchain, cloud computing, internet of things. That is the reason the product web design must include the significant attributes of the new technologies. Hence, blockchain technology needs to take advantage of traceability, cyber security, and communication. The analytical hierarchy process (AHP) decomposes a complicated problem into a multi-level hierarchical setup of criteria and characteristics, possessing decisive alternatives at the lowest level [64]. Application of AHP to determine relative importance for website element drivers is split into three stages: (i) the problem is decomposed into a top-down hierarchical structure, (ii) the weights for criteria evaluation are determined, and (iii) overall weight calculation for the website design element drivers. Comparison between criteria and a sub-criterion is carried out on the basis of the decision maker's/expert judgment evaluation, of which any of them is a critical concern to the website characteristic in the loop of the three echelon SC objectives.

In this research study, the hierarchical structure was considered three-level i.e., level 0 (objective) was web design index for the digital marketing of e-agricultural SCM, level 01 (criteria) was based on the blockchain technology index and basic index of web design, whereas level 02 (sub-criteria) considered the basic and blockchain-based WDIs. AHP was applied in the proposed work to determine the comparative weights for the SCM problem and rank the sub-criteria WDEs to evaluate the impact and significance on the objective, hence, pairwise comparison was carried out based on the data taken for each of the sub-criteria. The output of the AHP was the eigen vector, which expresses the impact or importance of the web design elements based on customers' and experts' opinion. Basic design and blockchain-based WDEs were then ranked based on their significance as given in Tables 2 and 3, respectively.

Table 2. The eigen vector for the evaluation of basic web design elements as an output of analytic hierarchy process (AHP).

Sr. No	Basic Web Design Elements	Eigen Values	Percentage	Ranking
1	Graphics and appearances (GA)	0.24	24.00	1
2	Card design (CD)	0.025	2.50	9
3	Hero images (HI)	0.052	5.20	7
4	Navigation (NA)	0.18	18.00	2
5	Web contents (WC)	0.085	8.50	5
6	Product videos (PV)	0.043	4.30	8
7	Typography (TY)	0.15	15.00	4
8	Branding (B)	0.071	7.10	6
9	Fast loading (FL)	0.16	16.00	3

Table 3. The eigen vector for the evaluation of blockchain elements as an output of AHP.

Sr. No	Blockchain Elements	Eigen Values	Percentage	Ranking
1	Search engine optimization (SEO)	0.24	24	1
2	Mobile compatibility (MC)	0.14	14	4
3	Machine learning (ML)	0.2	20	3
4	Email marketing (EM)	0.09	9	6
5	Enabled content management system (ECMS)	0.1	10	5
6	Cyber security (CS)	0.23	23	2

3.2. *Significant Web Design Elements using Pareto Analysis*

The Pareto analysis was applied to find the significant basic and blockchain-based WDEs for the promotion of the product and influencing customers towards products through digital marketing. The analysis is also called 80-20 rule, which is well illustrated by Figure 2. It was observed by the Pareto analysis that in case of blockchain, the effect of search engine optimization (SEO), cybersecurity (CS), and machine learning (ML) tools and the basic design elements i.e., web graphics (GA), navigation (NA), fast loading (FL), and typography (TY) are more than 80% effective in influencing customers in e-agricultural SCM. Few of these significant sub-criteria (WDIs) need to be explained for better understanding by the decision makers in digital marketing.

3.2.1. Web Graphics (GA)

Web graphics was ranked the most significant basic design elements among the screen out factors. Visual design is expressed as an eye-catching, esthetic, visual quality of a web page [65]. The fundamentals of visual design trade with emotional appeal, balance, esthetics, and homogeneity of the overall graphical look. We followed visuals such as layout/space, graphic improvisation, schematics, static/animations, and presentation of information which were identified by a comprehensive study of web designer's perception on Business-to-Corporate (B2C) websites [66]. It is evident from expert judgment that graphics and visuals are correlated with factors like making images appear swiftly and improving search engine index for quick discovery.

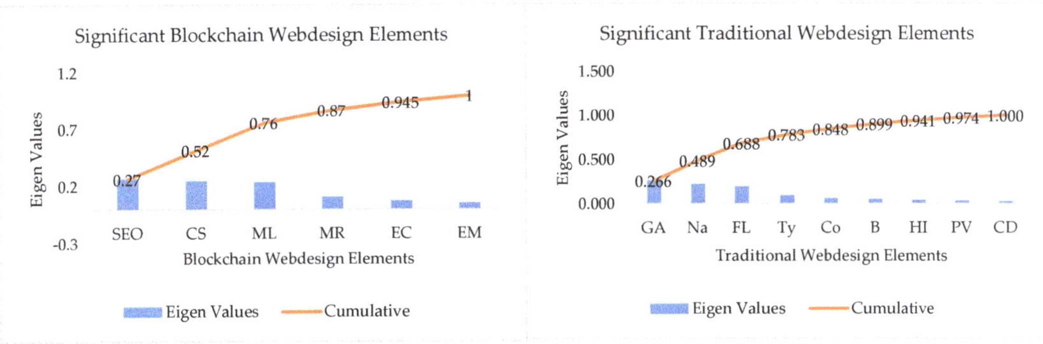

Figure 2. Pareto analysis (80-20 rule) for the selection of significant web design elements.

3.2.2. Search Engine Optimization (SEO)

SEO is an important tool of blockchain technology and is based on writing, coding, and designing a website in such a way that leads to an enhanced quality, volume, and visibility of a firm's website. It is phenomenal that around 90% of user visits are satisfactory from the results of a search engine, merely 5% of users avoid searching beyond the second page, and barely 2% move beyond the third page [67]. Appearance of a website on a foremost page is extremely competitive. In this work, internal website optimization was considered as an element of study. It included the design of website, keywords, and meta tags, which are essential for website itself, page names, links, photos, content texts on each page, and really-simple-syndication (RSS) feeds [68].

3.2.3. Cyber Security (SC)

Blockchain is based on security of the data and information. In our empirical study, cyber security was ranked second for website development consideration in blockchain. Several studies were empirically proven and presented by Wong et al. (previously referred [31]) that cybercrime, internet fraud, and identity theft affect the customer negatively, particularly in the aspect of online shopping. Trust is of paramount importance in the presence of uncertainty, because it mitigates information irregularity and assists customers to overcome the perception of risks. Hence, trust in electronic dealings is a significant differentiation that regulates any success or failure of firms. Further, parties are mostly unknown to each other during online transactions, so "initial trust" determines the occurrence of any transaction. Indeed, retailers' websites are a vital source of information to mitigate uncertainty. Therefore, cyber security is significant in order to sustain smooth operations on the electronic medium.

3.2.4. Fast Loading (FL)

Past studies presented a positive correlation between user satisfaction and website loading. Fast loading is extremely important for consumer willingness to conclude online transactions [69]. Speed of access can be measured indirectly via chronometer [70]. External factors facilitating the loading activity of a website include the applied hardware, web traffic, time of day, and others. Many researchers adopted the size presented in bytes of different homepages. In such a case, the heavier the homepage was, the more time was required to extract information inside the website. Furthermore, loading time of a page also represents quality of a website.

3.2.5. Navigation System (NG)

Navigation endorses structure design of a website. It refers to organization and accessibility of information displayed on the website [71]. While a user is operating on a website, the amount of effort is directly linked with the navigation design. Therefore, an effective navigation design must provide an effortless navigation hierarchy for users to access the desired information and its location on the website swiftly [72]. Navigation occupies the top relevant tier of significant factors because it plays a vital role in stimulating customers who are reluctant to be astounded by redundant links, screens, clicks, etc. Online customers want direct and simple navigation designs to save effort and time in searching for desirable items and completing their transaction in the smallest number of steps. An ambiguous navigation design irritates users and triggers them to lose location sense over the website; eventually they may leave the website without constructive activity. Previous studies recommend that an efficient navigation must assist users to traverse and navigate the site [73].

3.3. Fuzzy Inference System

To design the proposed fuzzy model, the combination of significant (three blockchain and four basic WDEs) web design elements were taken as an input of the fuzzy inference system (FIS), where the output was the web design index. Triangular membership function was applied to represent fuzzy input and output of the FIS system. The fuzzy numbers were defuzzified to crisp numbers using centroid method. The proposed model explicitly shows a mathematical function in which the image of n elements (n sub-criteria) is the final result of the model. Therefore, we can suppose the value y as a function f of n independent variables i.e.,

$$y = f(x1, x2, x3, \ldots, xn)$$

In the proposed FIS model, the number of inputs were considered as CS, SEO, ML, GA, NA, FL, and TY were considered from WDEs to find a web design index (WDI) as an output shown in the Figure 3.

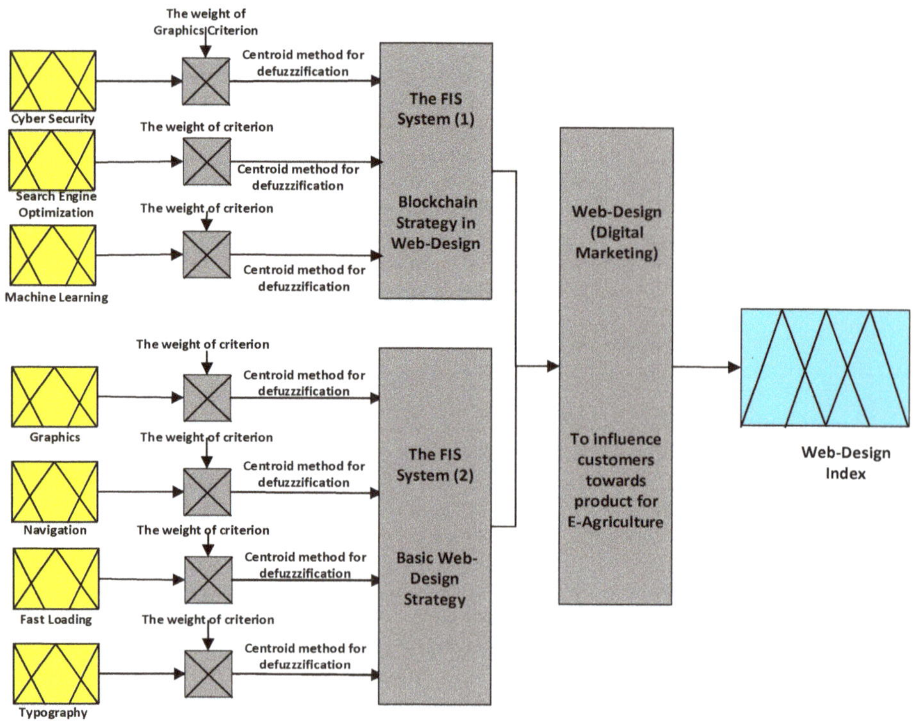

Figure 3. Flow diagram of fuzzy inference system (FIS) to represent input and output web design elements.

Each input (web design elements) and output (web design index) showed triangular fuzzy membership function having linguistic values i.e., "low", "medium", and" high". In FIS, low is represented by a value (0 1 2), whereas the medium and high are represented by (2 3 4) and (3 4 5) respectively. The rules were made based on the weights of each design elements as an input, which were taken from the eigen vector of the AHP. The centroid method was used to defuzzify the inputs into crisp value. Then, the obtained fuzzy numbers were defuzzified to the desired crisp numbers for using as input variables for the FIS systems in the first stage.

The output indices were generated using FIS by the combination of the significant input WDEs. This surface is well illustrated in the Figure 4 to show the relationship between the output (blockchain index and basic design index) and the input (significant WDEs). There were two FIS systems used in the process: one was for blockchain index and the other for basic design index, which were further combined for the calculation of web design index (WDI) though FIS. It is obvious that enhancing these input significant elements will result in increasing web design index and ultimately more customers will be attracted towards the product through digital marketing. This improvement will bear a cost of advertisement to be shared by the SCM partners for maximizing total profit by increasing e-market demand.

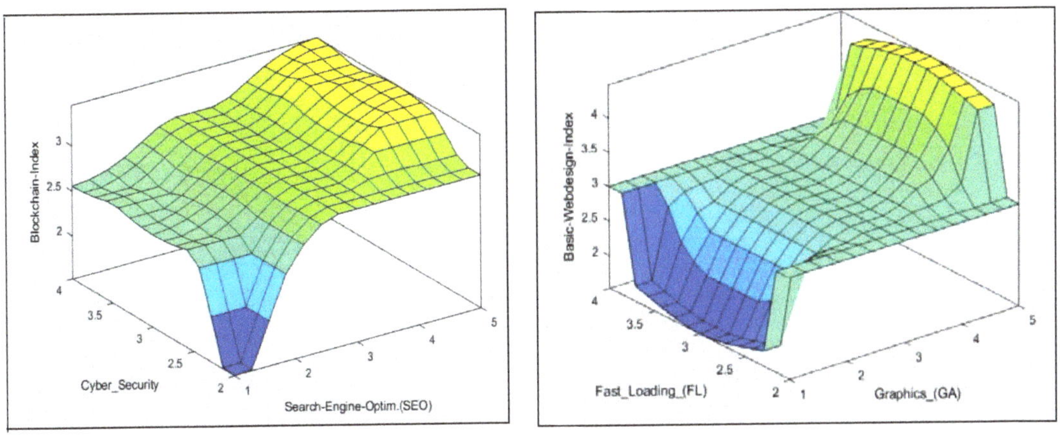

Figure 4. (**a**) The effect of web design elements on blockchain index. (**b**) The effect of web design elements on basic design index.

4. E-Agricultural Supply Chain Management

This section considers problem definition, e-advertisement, variable demand, fuzzy costs, notation, assumptions, and model formulation of the proposed co-op e-agricultural SCM model.

4.1. Problem Definition

The conventional agricultural supply chain management is extremely underdeveloped with respect to technology and innovation. Agri-firms are facing issues in communication for sharing information and products among SCM partners and are at a huge distance from the customers due to lack of technology. The agricultural product supply chain management (agri-SCM) requires an extensive development in communications and marketing strategy, which is possible by the addition of advanced technologies i.e., blockchain and e-agriculture (digitization). The research is based on digital marketing policy by enhancing web design elements for the promotion of the agri-product and influencing customers online to increase e-market demand in e-agri-SCM. A three-echelon supply chain model considering single supplier, single agri-processing firm, and multiple retailers are cooperating on the basis of e-advertising cost by various share agreements for maximum profit as shown in Figure 5. The uncertainties in the model were recovered by the fuzzing of all basic costs involved in calculating total profit of SCM.

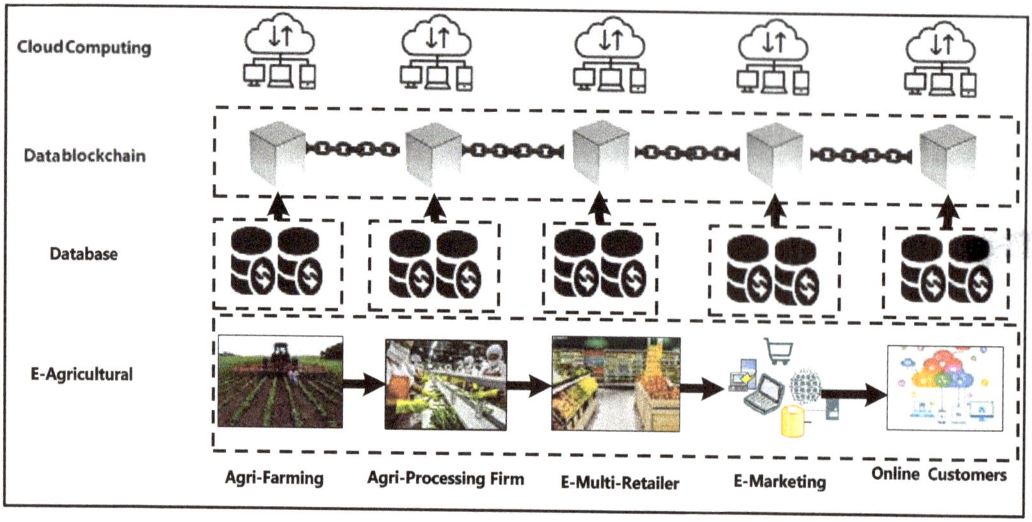

Figure 5. Flow diagram of cooperative e-agricultural supply chain management considering blockchain technology for digital marketing.

The blockchain technology is well processed through the e-agri-SCM, where the sharing of information, goods, and payments are made digitally using blockchain technology. The aim was to provide a developed e-agri-SCM by traceability, security, and communication system. All the data are stored in the database, integrated among SC players using blockchain, and then shared permanently by a cloud computing management system. The research provides a platform to the customers (web design) to provide basic design tools and blockchain tools for the security and traceability. The web design index was obtained from the significant WDEs incorporated in the product web design for the promotion and demand. The WDI was further linked with the variable demand of the agri-SCM to find the impact of the advancement on overall profit.

4.2. Online Advertisement Cost

A supply chain (SC) consisting single supplier, processing firm, and a multi-retailer was selected to analyze the e-advertisement co-op policy in agri-SCM. The proposed model turned out to be more realistic when we considered the cost of advertisement dependent on demand. Investment for the advertisement of products in the digital-market is backed by SC partners to enhance the average product sales for increasing demand from digital market as a result of increasing the e-advertisements efforts and vice versa. The demand in online marketing requires enhancing web design elements based on blockchain for the promotion of the product and it influences the customers for the increasing product demand. The blockchain tools can be incorporated in the product web design to ensure security and traceability among SC players and customers. The e- advertisement covers the investments to enhance the significant design elements on the product promotion web.

The online advertisement cost includes the expenses involved in designing the elements of the product's website. It is obvious that the advertisement cost depends on the website design index (WDI), i.e., Ω value starting from one to five. The expression shows a direct linear and exponential relationship between the advertisement cost and the website design elements as given in Equations (1) and (2), respectively. If the product web design is good for e-advertisement then it will have a WDI value almost 5 and there are maximum chances that more customers will be attracted towards the product. On the other hand, the bad or worst web design will compel the customers to review it as bad, which will ultimately affect the customer demand negatively in e-advertising strategy. The advertisement cost (AC) of the supply chain management is the sum of initial advertisement (a) cost and variable cost for online advertisement to enhance the website design for the attraction of the customers. i.e.,

$$AC = a + a(\Omega\lambda_1) \tag{1}$$

$$AC = a + e^{\lambda 2\,\Omega} \tag{2}$$

where a is the initial advertisement cost of the supply chain, Ω is the website design index, λa is the scaling factor of the linear curve, and λ_2 is the shape parameter in case of exponential variable advertisement cost. The above expressions are illustrated as in Figure 6.

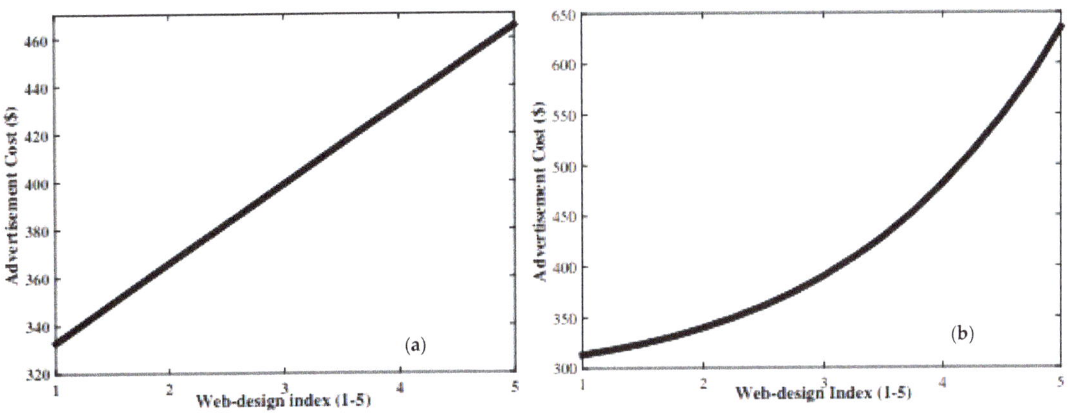

Figure 6. (**a**) The relationship between web design index and advertisement cost. (**b**) Exponential relationship between web design index and advertisement.

4.3. Model Notation

Notations used in the mathematical model of e-agri-SCM are as follows.

Indices

j index to represent number of retailers in SCM, j = 1, 2, . . . n

Decision Variables

T cycle time for retailers (year)
y_1 raw material taken by agri-farming, i.e., supplier (integer)
y_2 lots received by agri-processing firm (integer)
a_{rj} share of e-advertisement cost by jth retailer invested in digital marketing ($/year)
a_{mj} agri-processing firm's investment share for e-advertising the product ($/year)
a_{sj} supplier's investment in e-advertising agri-product for collaboration ($/year)
z_1 multiple units for the agri-firm's cycle time (integer)
z_2 retailer's cycle time multiple unit (integer)
p_j price of agri-product kept by the jth retailer in the market ($/unit)

Agri-farming/supplier's parameters

C_s processing cost of the semifinished agri-product ($/unit)
P_s processing rate of the semifinished agri-product (unit/year)
C_{fcs} cost of fixed carbon emission of the agri-farming/supplier ($/shipment/year)

C_{vcs} agri-farming/supplier's cost of variable carbon emission ($/unit/year)
$h_{s'}$ raw material holding cost ($/unit/year)
h_s agri-product holding cost of the supplier as a finishing product ($/unit/year)
A_s setup cost of the agri-farming/supplier ($/setup)
O_s cost of the agri-farming/supplier for ordering ($/order)
TC_s total cost of the agri-farming/supplier ($/year)

Agri-processing Parameters

P_m processing rate of semifinished products (unit/year)
h_m finished agri-product holding cost ($/unit/year)
C_{fcm} fixed cost as a tax of carbon emission by agri-processing firm ($/shipment/year)
C_{vcm} cost of variable carbon emission by agri-processing firm ($/unit/year)
A_m setup cost ($/setup)
O_m ordering cost ($/order)
F fixed/constant cost of transportation ($/shipment)
V agri-processing firm's varying transportation cost ($/shipment)
TC_m total cost of agri-processing firm ($/year)

Retailer's Parameters

h_r retailers holding cost per agri-product per cycle ($/unit/year)
A_r retailers ordering cost ($/order)
TC_r retailers' total business cost ($/year)
$D_{(p,\, a)}$ variable demand rate as a function of advertisement cost and selling price (units/year)

Other Parameters

Δ_2 upper limit of the fuzzy parameters (parameter units)
Δ_1 the lower limit of the membership function of the fuzzy number (parameter units)
d_j starting/initial demand of the jth retailer (units/year)
M total advertisement budget incorporated for the whole supply chain ($/year)
TC total cost of the e-agri-SCM ($/year)
TP the total profit achieved by the e-agri-SCM ($/year)

4.4. Demand Depending Selling Price and Advertising Cost

Demand is correlated with the selling price of the product and its advertisement cost. Indicating demand in terms of selling price and advertising cost as a decision variable was studied in the work of [74]. In the current case, previous function was altered to portray variability in demand owing to e-advertisement in digital-markets. The initial retail price and demand for the jth retailer were denoted as d_j and p_j, respectively. Both the demands, linear and exponential, were generated for the entire supply-chain once advertising cost investment in the digital-market was carried out, as depicted in Equations (3) and (4), respectively.

$$d_{rj} = d_j(\alpha - \beta p_j)^{\frac{1}{v}}(k_1(a_{sj} + a_{sj}(\Omega\lambda_1))^{\frac{1}{2}} + k_2(a_{mj} + a_{mj}(\Omega\lambda_1))^{\frac{1}{2}} + k_3(a_{rj} + a_{rj}(\Omega\lambda_1))^{\frac{1}{2}}) \tag{3}$$

$$d_{rj} = d_j(\alpha - \beta p_j)^{\frac{1}{v}}(k_1(a_{sj} + e^{\lambda_2\Omega})^{\frac{1}{2}} + k_2(a_{mj} + e^{\lambda_2\Omega})^{\frac{1}{2}} + k_3(a_{rj} + e^{\lambda_2\Omega})^{\frac{1}{2}}) \tag{4}$$

where α, β, and ν are non-negative constants. The values of v (i.e., $\nu < 1$, $\nu = 1$ or $\nu > 1$) is subject to the price demand dependency, which may follow a linear, concave, or convex curve separately. In this work, to exploit the supply chain profit, value of the ν was considered $\nu > 1$. Further, values of k_1, k_2, and k_3 are non-negative constants, which depicts the efficacy of local advertisement on behalf of the supplier, agri-processing firm, and vendors, separately in generating the demand from market. The overall demand of SC is produced from the local market from every single vendor/retailer, which is given in the following equation.

$$D = \sum_{j=1}^{n} d_{rj} \tag{5}$$

Likewise, the function shows the relationship amid demand, the advertising cost, and selling price as depicted in Figure 6. The portrayed surface shows the maximum bound of selling price, where the demand shrinks to zero, but does not illustrate the bound with aspect of advertisement cost, in other words, the demand enhances with the increase in advertisement cost without any limit/bound. Such behavior of limitation is regulated by a constraint studied in the proposed SC model.

4.5. Model Fuzzification for Uncertain Environment

Numerous uncertainty types exist in real life supply chain management problems. They are typically modeled via techniques from the probability theory. Nevertheless, some uncertainties persist that cannot be handled properly through conventional probabilistic models. Usually, the issue amounts to modeling SCM problems and finding their solutions under uncertain environment [75]. Hence, the way out is to treat such problems by fuzzy set theory instead of probability theory [76]. Centroid of fuzzy total cost was taken as the estimate for fuzzy total cost, but studies show that, in spite of centroid method, the signed distance method is better for defuzzification [77]. Let x be a fuzzy number with triangular membership function, $\widetilde{x} = (x - \Delta 1; x; x + \Delta 2)$, where $0 < \Delta_1 < x$ and $0 < \Delta_2 \leq 1 - x$. Further, Δ_1 and Δ_2 are determined by the decision makers. The x is given as in below Equation (6) from the study of [74].

$$d(\widetilde{x}, \widetilde{0}_1) = x + \frac{1}{4}(\Delta_2 - \Delta_1) \tag{6}$$

4.6. Model Assumptions

The e-agri-SCM model was developed based on the following assumptions.

1. An agricultural supply chain management model considering three partners for single type agricultural products [78,79]. The retailers are situated at different location and working online using digital marketing strategy of online web design including blockchain elements.
2. The basic costs of the e-agri-SCM are considered as fuzzy numbers to justify the uncertain environment [80].
3. Variable demand is considered depending advertisement cost and selling price, where the advertisement cost is the function of web design index, specifically $D(p, a) = a_o(\alpha - \beta p)^{\frac{1}{v}}(k\sqrt{a})$ [71].
4. The SC partners are collaborating on the basis of e-advertisment/online advertisement cost for maximizing total profit of the e-Agri-SCM.

4.7. SCM Model Formulation

Supply chain partners work together to enhance any product's demand. The agri-SCM model was formulated in such a way that identifies the overall basic along with miscellaneous costs that are associated to the supplier, agri-processor, and vendor/ retailer. The fundamental costs contained production cost, holding cost, setup cost, transportation cost, ordering costs, etc. With the advances in cleaner production for a sustainable environment and market growth, both the carbon emission cost and advertising cost were also integrated to construct the proposed SCM model to be further reliable and eco-efficient. Moreover, basic costs related to production and inventory are uncertain owing to various factors such as the world energy crisis, inflation, and oil prices. Due to this, all the fundamental costs linked with the supplier, agri-processing firm, and vendors/ retailers in the current proposed model were taken as fuzzy costs. The signed-distance formula was used to resolve the fuzzy-sets of parameters.

4.7.1. Supplier costs

The costs of supplier are as following.

Production Cost

A cost that is variable and depends on the production quantity for certain demand.

$$(C_s + \frac{1}{4}(\Delta_{cs2} - \Delta_{cs1}))D \tag{7}$$

Setup Cost

It involves the costs of initialization of an order from supplier perspective which is dependent on the cycle time and limitations set by the manufacturer or retailer.

$$\frac{(A_s + \frac{1}{4}(\Delta_{as2} - \Delta_{as1}))}{z_1 z_2 T} \tag{8}$$

Ordering cost

It includes costs incurred to order a batch or purchase of externally made products that is dependent on the quantity limits set by the manufacturer or retailer.

$$\frac{(O_s + \frac{1}{4}(\Delta_{os2} - \Delta_{os1}))y_1}{z_1 z_2 T} \tag{9}$$

Holding cost of raw material

This is the cost that is incurred while holding inventory or stocking in a storage or warehouse that is dependent upon parameters like cycle time, variable demand, and production of the products.

$$(h_{s'} + \frac{1}{4}(\Delta_{hs'2} - \Delta_{hs'1}))\frac{z_1 z_2 T D^2}{2y_1 P_s} \tag{10}$$

Holding cost of finished products

The cost of the holding was taken from the work of Sarkar et al. [81].

$$\frac{(h_s + \frac{1}{4}(\Delta_{hs2} - \Delta_{hs1}))z_2 T}{2}[(\frac{2}{y_2} - z_1)\frac{D^2}{P_s} + (1 - \frac{1}{y_2})\frac{D^2}{P_m} + (z_1 - 1)D] \tag{11}$$

E-advertisement cost

The supplier is sharing an advertisement cost for the enhancement of web design by incorporating blockchain and basic elements to influence customers towards agri-products based on co-op e-advertising collaboration policy.

$$\sum_{j=1}^{n} \frac{a_{sj} + a_{sj}(\Omega\lambda_1)}{z_1 z_2 T} = \frac{(a_{s1} + a_{s2} + a_{s3} \ldots + a_{sn})(1 + (\Omega\lambda_1))}{z_1 z_2 T} \tag{12}$$

Carbon emissions cost

The aim of the incorporating carbon cost was to provide the necessity of present and future generations for the sustainable and cleaner supply chain management.

$$C_{fcs} y_1 + C_{vcs} D \tag{13}$$

The total cost of the supplier (farming industry) in agri-SCM can be expressed mathematically as

$$\widetilde{T}C_s = (C_s + \tfrac{1}{4}(\Delta_{cs2} - \Delta_{cs1}))D + \tfrac{(A_s+\frac{1}{4}(\Delta_{as2}-\Delta_{as1}))+(O_s+\frac{1}{4}(\Delta_{os2}-\Delta_{os1}))y_1}{z_1 z_2 T}$$
$$+(h_{s'} + \tfrac{1}{4}(\Delta_{hs'2} - \Delta_{hs'1}))\tfrac{z_1 z_2 T D^2}{2 y_1 P_s} + (h_s + \tfrac{1}{4}(\Delta_{hs2} - \Delta_{hs1}))\tfrac{z_2 T}{2}((\tfrac{2}{y_2} - z_1)\tfrac{D^2}{P_s} \tag{14}$$
$$+(1 - \tfrac{1}{y_2})\tfrac{D^2}{P_m} + (z_1 - 1)D) + \sum_{j=1}^{n}(\tfrac{a_{sj}+a_{sj}(\Omega\lambda_1)}{z_1 z_2 T}) + (C_{fcs} y_1 + C_{vcs} D).$$

4.7.2. Agri-Processing Cost

The agri-processing firm processes the finished product of the supplier (farming industry) to produce the agri-product. The basic costs associated with the agri-processing firm are considered to be fuzzy due to uncertain conditions, i.e., processing, setup, ordering, and holding costs. Other specific costs are also incurred by the firm in the form of transportation and carbon emission costs.

Processing cost

This is a cost that is variable and depends on the agri-processing cost of the products and their quantity for certain demand.

$$(C_m + \frac{1}{4}(\Delta_{cm2} - \Delta_{cm1}))D \tag{15}$$

Setup cost

It involves costs of agri-processing setup in terms of material and labor required to make the machine ready for new production lot.

$$\frac{(A_m + \frac{1}{4}(\Delta_{am2} - \Delta_{am1}))}{z_2 T} \tag{16}$$

Ordering cost

It includes costs incurred to order an agri-based batch or purchase of externally made agri-products that is dependent on the quantity limits set by the manufacturer or retailer.

$$\frac{(O_m + \frac{1}{4}(\Delta_{om2} - \Delta_{om1}))y_2}{z_2 T} \tag{17}$$

Holding cost of raw material

Cost dealing with holding of agri-environment based materials.

$$\frac{(h_s + \frac{1}{4}(\Delta_{hs2} - \Delta_{hs1}))Q_m D}{2P_m} \tag{18}$$

Holding cost of finished product

The fuzzy holding cost can be expressed as (previously referred Sarkar et al. [77])

$$\frac{(h_m + \frac{1}{4}(\Delta_{hm2} - \Delta_{hm1}))T}{2}[(2 - z_2)\frac{D^2}{P_m} + (z_2 - 1)D] \tag{19}$$

E-advertisement cost

Similarly, agri-processing firms also invest a share based on the e-advertisement collaboration policy for the increasing demand in e-marking using digital marketing by incorporating web design elements. The agri-processing firm's share can be expressed mathematically as

$$\sum_{j=1}^{n}\frac{a_{mj}+a_{mj}(\Omega\lambda_1)}{z_2T}=\frac{(a_{m1}+a_{m2}+a_{m3}\ldots+a_{mn})(1+(\Omega\lambda_1))}{z_2T}. \quad (20)$$

Carbon emission cost

$$C_{fcm}y_2+C_{vcm}D \quad (21)$$

The total cost of the agri-processing firm under uncertain conditions is given as

$$\begin{aligned}\widetilde{T}C_m &= (C_m+\tfrac{1}{4}(\Delta_{cm2}-\Delta_{cm1}))D+\tfrac{(A_m+\frac{1}{4}(\Delta_{am2}-\Delta_{am1}))+(O_m+\frac{1}{4}(\Delta_{om2}-\Delta_{om1}))y_2}{z_2T}\\ &+(h_s+\tfrac{1}{4}(\Delta_{hs2}-\Delta_{hs1}))\tfrac{z_2TD^2}{2y_2P_m}+(h_m+\tfrac{1}{4}(\Delta_{hm2}-\Delta_{hm1}))\tfrac{T}{2}((2-z_2)\tfrac{D^2}{P_m}+(z_2-1)D)\\ &+\sum_{j=1}^{n}(\tfrac{a_{mj}+a_{mj}(\Omega\lambda_1)}{z_1z_2T})+\tfrac{y_2F+VQ_m}{z_2T}+(C_{fcm}y_2+C_{vcm}D).\end{aligned} \quad (22)$$

4.7.3. Retailers' Costs

The retailers in digital businesses are mainly concerned with the warehousing and they are keeping the agri-product of the SCM. The agri-products are deteriorated items, which are also placed on the website with basic and blockchain features. There are multiple e-retailers connected to the agri-processing firm. The agri-processing firm ships the required agri-products to the retailers and big food malls based on the customers' demand online and offline. The retailers' costs can be divided into the following sub costs.

Ordering cost

$$\frac{(O_r+\frac{1}{4}(\Delta_{or2}-\Delta_{or1}))}{T} \quad (23)$$

Holding cost of raw material

The holding cost of all jth retailers in the agri-SCM is expressed as

$$\frac{(h_r+\frac{1}{4}(\Delta_{hr2}-\Delta_{hr1}))TD}{2}. \quad (24)$$

Advertisement cost

Retailers also share their role in the collaboration policy by investing in e-advertisement through digital marketing. Their integrated sum can be expressed as

$$\sum_{j}^{n}\frac{a_{rj}+a_{rj}(\Omega\lambda_1)}{T}=\frac{(a_{r1}+a_{r2}+a_{r3}\ldots+a_{rn})(1+(\Omega\lambda_1))}{T}. \quad (25)$$

Hence, the total cost of the retailers in Agri-SCM business is given as

$$\widetilde{T}C_r=\frac{(O_r+\frac{1}{4}(\Delta_{or2}-\Delta_{or1}))}{T}+(h_r+\frac{1}{4}(\Delta_{hr2}-\Delta_{hr1}))\frac{TD}{2}+\sum_{j}^{n}\frac{a_{rj}+a_{rj}(\Omega\lambda_1)}{T}. \quad (26)$$

4.7.4. Total Cost of the Supply Chain

After formulating the cost of supplier (farming), cost of agri-processing firm, and cost of retailers in dealing agri-product, the total cost of the e-agri-SCM can be expressed mathematically as

$$\widetilde{T}C=\widetilde{T}C_s+\widetilde{T}C_m+\widetilde{T}C_r. \quad (27)$$

By substituting $\widetilde{T}Cs$, $\widetilde{T}C_m$, and $\widetilde{T}C_r$ the total fuzzy cost of the supply chain is given as in Equation (30).

$$\widetilde{T}C = (C_s + \tfrac{1}{4}(\Delta_{cs2} - \Delta_{cs1}))D + \frac{(A_s+\frac{1}{4}(\Delta_{as2}-\Delta_{as1}))+(O_s+\frac{1}{4}(\Delta_{os2}-\Delta_{os1}))y_1}{z_1 z_2 T} + \frac{z_1 z_2 T D^2}{2 y_1 P_s}$$
$$(h_{s'} + \tfrac{1}{4}(\Delta_{hs'2} - \Delta_{hs'1})) + \tfrac{z_2 T}{2}(h_s + \tfrac{1}{4}(\Delta_{hs2} - \Delta_{hs1}))((\tfrac{2}{y_2} - z_1)\tfrac{D^2}{P_s} + (1 - \tfrac{1}{y_2})\tfrac{D^2}{P_m} + (z_1 - 1)D)$$
$$+ \sum_{j=1}^{n}(\frac{a_{sj}+a_{sj}(\Omega\lambda_1)}{z_1 z_2 T}) + \frac{y_1 F + V Q_s}{z_1 z_2 T} + (C_{fcs} y_1 + C_{vcs} D) + (C_m + \tfrac{1}{4}(\Delta_{cm2} - \Delta_{cm1}))D + \frac{z_2 T D^2}{2 y_2 P_m}$$
$$(h_s + \tfrac{1}{4}(\Delta_{hs2} - \Delta_{hs1})) + \frac{(A_m+\frac{1}{4}(\Delta_{am2}-\Delta_{am1}))}{z_2 T} + \frac{(O_m+\frac{1}{4}(\Delta_{om2}-\Delta_{om1}))y_2}{z_2 T} \tag{28}$$
$$+ \tfrac{T}{2}(h_m + \tfrac{1}{4}(\Delta_{hm2} - \Delta_{hm1}))((2 - z_2)\tfrac{D^2}{P_m} + (z_2 - 1)D) + \sum_{j=1}^{n}(\frac{a_{mj}+a_{mj}(\Omega\lambda_1)}{z_1 z_2 T}) + \frac{y_2 F + V Q_m}{z_2 T}$$
$$+ (C_{fcm} y_2 + C_{vcm} D) + \frac{(O_r+\frac{1}{4}(\Delta_{or2}-\Delta_{or1}))}{T} + \tfrac{TD}{2}(h_r + \tfrac{1}{4}(\Delta_{hr2} - \Delta_{hr1})) + \sum_{j}^{n}\frac{a_{rj}+a_{rj}(\Omega\lambda_1)}{T}$$

The firms' revenue formula can be written as.

$$Revenue = \sum_{j=1}^{n}(p_j d_{rj}) \tag{29}$$

and the profit formula is

$$Totalprofit(TP) = Revenue - TC. \tag{30}$$

Now, the objective of the proposed model is to maximize the total profit of e-agri-SCM, which is given as

$$Maximize\widetilde{T}P = \sum_{j=1}^{n}(p_j d_{rj}) - [(C_s + \tfrac{1}{4}(\Delta_{cs2} - \Delta_{cs1}))D + \frac{(A_s+\frac{1}{4}(\Delta_{as2}-\Delta_{as1}))}{z_1 z_2 T} \times$$
$$+ \frac{(O_s+\frac{1}{4}(\Delta_{os2}-\Delta_{os1}))y_1}{z_1 z_2 T} + \frac{z_1 z_2 T D^2}{2 y_1 P_s}(h_{s'} + \tfrac{1}{4}(\Delta_{hs'2} - \Delta_{hs'1})) + \tfrac{z_2 T}{2}(h_s + \tfrac{1}{4}(\Delta_{hs2} - \Delta_{hs1}))$$
$$((\tfrac{2}{y_2} - z_1)\tfrac{D^2}{P_s} + (1 - \tfrac{1}{y_2})\tfrac{D^2}{P_m} + (z_1 - 1)D) + \sum_{j=1}^{n}(\frac{a_{sj}+a_{sj}(\Omega\lambda_1)}{z_1 z_2 T}) + \frac{y_1 F + V Q_s}{z_1 z_2 T}$$
$$+ (C_{fcs} y_1 + C_{vcs} D) + (C_m + \tfrac{1}{4}(\Delta_{cm2} - \Delta_{cm1}))D + \frac{(A_m+\frac{1}{4}(\Delta_{am2}-\Delta_{am1}))}{z_2 T} \tag{31}$$
$$+ \frac{(O_m+\frac{1}{4}(\Delta_{om2}-\Delta_{om1}))y_2}{z_2 T} + \frac{z_2 T D^2}{2 y_2 P_m}(h_s + \tfrac{1}{4}(\Delta_{hs2} - \Delta_{hs1}) + \tfrac{T}{2}(h_m + \tfrac{1}{4}(\Delta_{hm2} - \Delta_{hm1}))$$
$$((2 - z_2)\tfrac{D^2}{P_m} + (z_2 - 1)D) + \sum_{j=1}^{n}(\frac{a_{mj}+a_{mj}(\Omega\lambda_1)}{z_1 z_2 T}) + \frac{y_2 F + V Q_m}{z_2 T} + (C_{fcm} y_2 + C_{vcm} D)$$
$$+ \frac{(O_r+\frac{1}{4}(\Delta_{os2}-\Delta_{os1}))}{T} + \tfrac{TD}{2}(h_r + \tfrac{1}{4}(\Delta_{hr2} - \Delta_{hr1})) + \sum_{j}^{n}\frac{a_{rj}+a_{rj}(\Omega\lambda_1)}{T}.$$

The constraints/limitations of the proposed SCM model are given below.

Advertisement costs constraints

$$\sum_{j=1}^{n}(a_{sj} + a_{mj} + a_{rj}) \leq M \tag{32}$$

$$\sum_{j=1}^{n} a_{sj} \leq M_s \tag{33}$$

$$\sum_{j=1}^{n} a_{mj} \leq M_m \tag{34}$$

$$\sum_{j=1}^{n} a_{rj} \leq M_r \tag{35}$$

Space constraints

$$\frac{(P_s - D)(DTz_1z_2)}{P_s} \leq I_s \tag{36}$$

$$\frac{(P_m - D)(DTz_2)}{P_m} \leq I_m \tag{37}$$

$$Td_{rj} \leq I_{rj}$$

5. Numerical Experiment

The intended agricultural supply chain management needs to be validated pragmatically by numerical analysis of a real-life example. A fundamental three-echelon SCM having a single supplier and agri-processing firm each and three-retailers, i.e., (j = 3), was considered. All the partners collaborated through a co-op e-advertisement to invest in enhancing a website for agri-product marketing. The numerical experiment of the research study was based on agri-SCM considering a sugar processing firm, its suppliers, and retailers. The data utilized to perform the experiment were taken from the local industry of sugar processing SCM. The suppliers/agri-farming data are given in Table 4 consisting of production, holding, transportation, carbon emission, and production rate. The capacity of the sugar processing firm, inventory space, and budget data were also taken from the selection industry to deal with the constraints of the proposed model. The data were reliable and provided a pragmatic application of the proposed e-agri-SCM mathematical model with blockchain technology to attract customers. Principally, the three-echelon agri-SCM starts with the farming industry (supplier). The data for processing cost, setup cost, holding cost, transportation cost, and carbon emission are given in Table 4. The web design index (WDI) was taken as $\Omega = 3$ for both numerical experiments and sensitivity analysis.

Table 4. Suppliers/agri-farming cost data ($).

Δ_{cs1}	C_s	Δ_{cs2}	Δ_{as1}	A_s	Δ_{as2}	Δ_{hs1}	h_s	Δ_{hs2}	Δ_{os1}	O_s	Δ_{os2}	$\Delta_{hs'1}$	hs'	$\Delta hs'2$	P_s	C_{fcs}	C_{vcs}
1	2	3	450	500	550	0.45	0.6	0.75	275	300	325	0.3	0.4	0.5	299	0.2	0.1

The agri-processing firm data are also crucial for the analysis of the proposed model due to the important stake in the chain. Further, the agri-processing firm is also an imperative supply chain partner in co-op advertisement collaboration. The cost data associated with the agri-processing firm's (sugar processing firm) production, inventory, carbon emission costs, and transportation are presented in Table 5. All these costs are uncertain and therefore reflected with fuzzy parameters.

Table 5. Agri-processing firm's cost data ($).

Δ_{cm1}	C_m	Δ_{cm2}	Δ_{am1}	A_m	Δ_{am2}	Δ_{hm1}	h_m	Δ_{hm2}	Δ_{om1}	O_m	Δ_{om2}	F	V	P_m	C_{fcm}	C_{fcm}
2	3	4	180	200	220	4	5	6	135	150	165	0.2	0.1	1900	0.1	0.1

The role of retailers in e-agri-SCM for sugar production is very influential to deal with the customers. The data for retailers are given in Table 6.

Further, inventory capacities of the sugar-processing firm, supplier, and each retailer were subject to constraints. Similarly, a limitation cap was also set on the sharing of advertisement cost among the SCM partners spending the complete advertisement budget. The data relevant to the constraints of the proposed model are shown in Table 7.

Table 6. Data related to retailers.

Retailer Type	Initial Demand (units)	Δ_{hr1}	Holding Cost ($/unit/year)	Δ_{hr2}	Δ_{or1}	Ordering Cost ($/Order)	Δ_{or2}
R1	100	1.33	1.66	2	6.66	10	13.33
R2	150	1.33	1.66	2	6.66	10	13.33
R3	130	1.33	1.66	2	6.66	10	13.33

Table 7. Limitations of inventory and advertisement cost.

Advertisement Budget	Supplier Capacity	Processing Firm Capacity	Retailer-1 Inventory	Retailer-2 Inventory	Retailer-3 Inventory
3000	500	300	180	220	200

Next, the data were analyzed by application of non-linear optimization approach. Both the selling price of the product and the advertisement costs among supply chain partners were optimized under a co-op advertising collaboration policy.

6. Solution Methodology and Result Findings

The role of e-advertising and digital marketing is essential to achieve competitive advantage among production firms and is more highlighted in processing agricultural products. The suggested SCM model is based on a non-linear maximization problem alongside multiple variables and constraints. Previously, the constrained problem was resolved and analyzed by converting it into an unconstrained problem, exploring and exploiting for the global optimal solution. However, these approaches have now been discovered to be comparatively ineffective and have been substituted by approaches based on the equations of Karush–Kuhn–Tucker (KKT).

KKT equations possess the required conditions for optimizing a constrained problem. Such equations also deliver a solution to several nonlinear problems by means of computing Lagrange multipliers directly. Further, these approaches are termed as sequential quadratic programming (SQP) methods [82,83].

The set of equations created from the proposed model consist of non-linear equations sufficiently complex to solve by any analytical technique. Nevertheless, these analytical techniques are also ineffective and time-consuming and have been substituted by the techniques constructed on quadratic programming. The SQP is an effective and efficient decision-making tool and validated the best for solving nonlinear constraint and unconstraint equations, big-data research, and multi-decision problems according to Schittkowski [84], Mostafa and Khajavi [85], and Theodorakatos et al. [86]. The method is already utilized by various research studies [87–90] in production and supply chain management models.

There are five different cases considered from the real-life studies for the application of the proposed SCM by incorporating co-op advertising policy and web design index for increasing demand and influencing customers towards product. These cases are explained in detail in the following. The decision variables in terms of cycle time, shipments, selling prices, and corresponding advertisement costs for the agri-processing firm, supplier and retailers are given in Table 8.

6.1. Case 01

This scenario reflects the proposed cooperative SCM by considering variable advertisement cost as a function of the web design index (WDI), which is well illustrated in Figure 6, since total advertisement cost is the sum of initial cost (traditional) and variable cost depending on the WDI. The latter is different at different locations (e.g., developing and developed countries, urban and rural, cities and villages, etc.) This is the only case which considers the advertisement cost as an exponential variant of the WDI. This case has no restriction over e-advertisement cooperation share among SC partners. The analysis of the sequential quadratic programming (SQP) showed total profit of SCM as $827,049.16, with the optimal cycle time, shipments, selling prices, and advertisement shares by the SCM participants. It was observed that selling prices ($454.8) and shipment sizes (y1 = 1 and y2 = 2) were the same in all cases. The advertisement costs shared by the SCM participants varied depending on the agreement, leader-follower game theory, and advertisement budget constraint.

Table 8. Solution obtained by the analysis of numerical experiment using the sequential quadratic programming (SQP) method with $\Omega = 3$.

Sr. No.	Decisions	Symbol	Case 01	Case 02	Case 03	Case 04	Case 05
1	Cycle time (year)	T_j	0.225	0.227	0.233	0.237	0.209
2	Multiple (integer)	z_1	1	1	1	1	1
3		z_2	4	4	3	3	4
4	Shipments (number)	y_1	1	1	1	1	1
5		y_2	2	2	2	2	2
6	Selling price ($)	p_1	454.88	454.798	454.88	454.89	454.86
7		p_2	454.88	454.79	454.88	454.89	454.86
8		p_3	454.88	454.798	454.886	454.89	454.86
9	Advertisement investments in e-markets ($)	a_{r1}	190.920	196.782	399.281	416.66	312.5
10		a_{r2}	448.17	442.76	313.81	320	240
11		a_{r3}	332.922	332.56	256.57	263.333	197.5
12		a_{m1}	200.43	206.12	425.35	416.66	625
13		a_{m2}	469.57	463.78	325.69	320	480
14		a_{m3}	348.99	348.35	266.71	263.33	395
15		a_{s1}	200.432	206.125	425.35	416.6	312.5
16		a_{s2}	459.55	455.15	320.48	320	240
17		a_{s3}	348.99	348.35	266.71	263.33	197.5
18	Total profit	TP	827,049.16	856,239.90	832,242.86	832,195.13	820,931.7012

6.2. Case 02

This case is applicable when the advertisement cost is a linear function of the WDI to attract customers. There is no restriction of the e-advertisement budget by the supply chain participants. In addition, the scenario does not reflect superior or leader/follower SCM. It was observed that the total profit of the SCM obtained as highest among all cases as $856,239.9, where the cycle time was almost 0.227 years and the advertisement costs showed the global optimal values for supplier, agri-processing firm, and multi-retailers in SCM without restriction.

6.3. Case 03

This case reflects the scenario where there is a limitation of the total e-advertisement budget as given in Equation (33) provided by the unequal share of the supplier, agri-processing firm, and retailer. The SCM participants will invest in total e-advertisement within the total limit of the budget for increasing demand. Here again the advertisement cost is taken as the linear function of WDI. Here, the total profit obtained as $832,242.86 was slightly more than Case 1 but comparatively much lower than Case 2.

6.4. Case 04

This case represents the real-life scenario where there is a limitation on the budget of each players of the SCM. Irrespective of Case 03, in this case each participant is restricted to share under the limited budget as expressed by the Equations (34) and (35). This co-op e-advertising policy is undertaken by the optimal equal share of the participants for the promotion of product in various e-markets. The results showed a slight decrease in total profit, i.e., $832,195.13, as compared to the Case 03.

6.5. Case 05

The final case shows the policy of superior supply chain management, where the agri-processing firm is plays the role of a leader in the SCM whereas the supplier and multi-retailer are the followers. In three-echelon supply chain management, the processing is probably on a leading side as compared to supplier and retailers to handle all the main activities of the product life cycle. That is the reason a high profit is expected by the agri-processing firm in supply chain management in case of superior co-op advertisement policy. Agri-processing shares 50%, while the remainder is equally divided by the supplier and multi-retailer for the maximization of the total profit. This case showed the total profit of the three echelon as the lowest among all cases i.e., $82,0931.7.

The optimal solutions from the proposed five scenarios are valuable to the decision-makers for prediction and understanding of the uncertain/variable demand produced by e-marketing via web design index. Furthermore, these results deliver the optimal mode of distributing advertisement expenses among the supply chain partners through a policy of co-op e-advertising collaboration. This policy relies on a common mode of e-marketing, where each supply-chain partner presents the cost on enhancing the web design to advertise agri-products and surge the demand of each vendor/retailer by feedback from the consumers. Consequently, the demand from each vendor/retailer puts pressure on the agri-processing firm as well, thus the supplier is pushed to increase the production, and eventually a high profit is expected. The proposed model is nonlinear in nature and generic to help the decision makers in calculating multi-dimensional variables, i.e., shipment size, cycle time, advertisement share, and selling price of product. The model can be applied to firms and SCM, where the point of interest is to get maximum profit by the collaboration of supplier, manufacturer, and multi-retailer. The proposed model deals with the design of the product website by incorporating advanced elements (WDEs), which are commonly available on various products' website and justified. The research study shows a practical implication of the proposed model in the agricultural SCM to manage the supply chain network. The numerical experiments are based on the data collected from the agricultural processing firm and its supply chain partners for the pragmatic application of the research. The agricultural firms can take advantages of the digital marketing and blockchain to enhance the agri-product sale and to attain maximum profit. One feature of the proposed model is to provide room for a collaborative policy implementation among the supply chain partners to agree on the exact amount of advertising investment required to enhance profit for the whole supply-chain.

7. Sensitivity Analysis

The proposed SQP decision tool is effective in analyzing nonlinear optimization problem. Various scenarios have been analyzed and developed in the sensitivity analysis of the proposed e-agri-SCM for the pragmatic results. The demand in online marketing requires e-advertisement to enhancing design elements of the web for the promotion of the product and it influences the customers for the increasing product demand. The e-advertisement covers the investments to enhance the significant design elements on the product promotion web. In this regard, a detailed sensitivity analysis is essential to find the impact of the linear and exponential web design index value obtained from the FIS system of the selected significant web design elements on total demand, each e-market demand, cycle time, and total profit of SCM, which is given in Table 9.

1. It is observed that the total demand of the product and total profit of the SCM increases linearly by increasing the WDI (Ω). The observation is clearer by illustrating the left-hand-side curve of Figure 7. It is clear that by increasing WDI from level 01 to 05, the total demand of the product increases from 1435 to 1542 items whereas the total profit increases from $824,999 to $886,448.
2. On the other hand, by observing the exponential analysis, the total demand and the centralized total profit obtained by the participants increases exponentially as shown in the right-hand-side curve of Figure 8. It was observed that the total demand and

total profit also increase with respect to the WDI levels. However, in this case the impact is more because both demand and total profit have increased from 1410 units and $811,879 to 1587 units and $913,388.61 by investing to enhance the WDI level from 1 to 4 respectively.

These results provide an essential insight for managers to understand the importance of web design in digital marketing to consider the significant elements. These insights support the economic benefit of digital marketing in a co-op advertising policy to capture local and international market in three-echelon SCM by maximizing total profit.

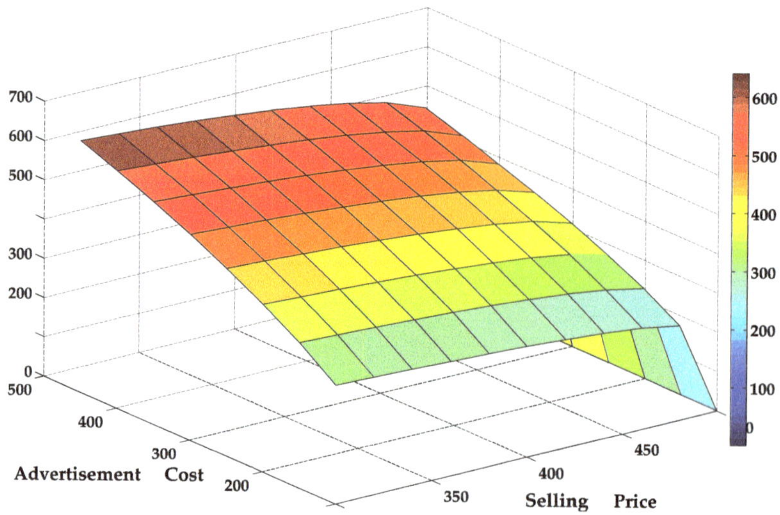

Figure 7. Representation of demand based on selling price and advertisement cost.

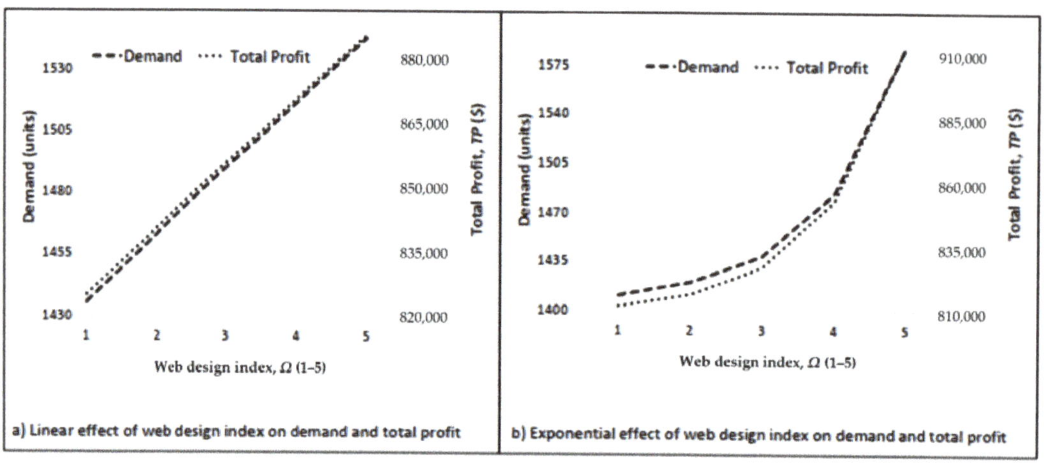

Figure 8. The effect of web design index on increasing customer demand and total profit of supply chain management.

Table 9. Evaluation of Demand and Cycle times for established parameters based on WDI.

Parameters.	Index (WDI)	Demand Units				Cycle Time	Total Profit
		dr1	*dr2*	*dr3*	*D*	*Tj* (years)	*TP* ($)
Web design elements (Linear)	Ω = 1	378.29	565.53	491.78	1435.60	0.23	824,999.79
	Ω = 2	385.52	576.40	501.17	1463.09	0.23	840,756.93
	Ω = 3	392.62	587.07	510.40	1490.09	0.23	856,239.90
	Ω = 4	399.59	597.57	519.47	1516.63	0.23	871,465.18
	Ω = 5	406.46	607.90	528.39	1542.75	0.23	886,448.01
Web designel ements (Exponential)	Ω = 1	371.81	558.26	480.92	1410.98	0.23	811,879.84
	Ω = 2	374.30	559.49	486.59	1420.38	0.23	816,354.00
	Ω = 3	379.16	566.76	492.91	1438.84	0.22	827,049.16
	Ω = 4	390.88	584.27	508.15	1483.30	0.22	852,824.13
	Ω = 5	418.45	625.23	543.98	1587.65	0.21	913,388.61

8. Conclusions

The development in e-businesses over the past decades has significantly strengthened the supply chain management (SCM) and given firms a competitive marketing opportunity. Agri-firms are searching for smart technologies to sell their product and to convince their customers with DM-SCM. Digitization considering blockchain effect is essential for the development of a sustainable agri-SCM due to the ease in product traceability, security, ease in transactions, etc. This paper presented a successful co-op e-advertisement collaboration policy among the supplier, agri-processing firm, and retailers in sustainable agricultural product supply chain management (SCM) by enhancing product web design incorporating the blockchain effect to attract customers. Product demand was considered as a variable depending on selling price and advertising cost. Web design elements were filtered to significant ones and later on used to estimate web design index (WDI), which was incorporated in the model against uncertain demand. Further, advertising cost was also considered as a function of the WDI, which was obtained by processing significant web design elements (WDEs) by the combination of the factors of the blockchain and basic design. Analytically hierarchical process (AHP), Pareto analysis, and fuzzy inference system (FIS) were used to identify and select the significant WDIs. The WDI was further incorporated into the mathematical model of the supply chain management (SCM) and calculated a positive effect of enhancing web design elements on the total profit. The results were outstanding in the form of optimal selling prices, shipments, cycle time, and optimal advertisement costs.

The model provides supports the decision-makers for keeping the best-selling prices and optimal e- advertisement share supported by supply chain participants to maximize the total profit of the Agri-SCM. Five different cases were taken based on the e-advertisement budget, partner's share, and demand curve for the application of the proposed model in real-life industrial problems. It was found that Case 02 shows a maximum profit, where the partners are required to share an optimal e-advertisement cost for digital marketing. The equal share cooperation policy based on advertisement cost exhibited by the participants was reflected in Case 04 for the digital marketing of the product. The results of superior SCM (given in Case 05), where the agri-processing firm is the leader and other participants are the followers showed the lowest profit of SCM. In addition, the research model is also beneficial to the customers by providing an interface of advance digital marketing strategy coupled with blockchain effect for traceability, security, and on time agri-product delivery. The selling prices was also less as compared to the traditional marketing agri-products due to fewer transportation costs and third-party logistics costs. These outstanding results and optimal solutions are very important for the industrial managers and decision-makers to successfully attain global supply chain management through proposed cooperative e-advertising policy.

The uncertainties in the proposed model were justified using fuzzy systems. The basic costs of the SCM mathematical model were considered as triangular fuzzy and formulated by using a signed distance method. Similarly, the identification of significant WDEs, i.e., web graphics, search engine optimization, cyber-security, fast loading, and navigation, is essential for digital marketing to convince customers towards the product in global SCM. These WDEs were found to be intangible and have been processed through the centroid method of FIS. These identified significant WDEs can be utilized for the future research and extended research of digital marketing. Moreover, the model incorporated carbon emission cost for providing cleaner environment and eco-effective production at supply chain level. The proposed mathematical model can be extended towards stochastic by considering expected demand and other important parameters of the SCM. The cooperation can be based on the whole sale price or selling price of the product. Overall, the research is a significant contribution by specifically targeting web design in the field of digital marketing for global SCM.

Author Contributions: Conceptualization, M.A. and M.O.; methodology, M.A.; software, Q.S.K.; validation, M.A. and Q.S.K.; formal analysis, M.A., M.J., and G.H.; investigation, G.H. and C.I.P.; data curation, M.O. and G.H.; writing—original draft preparation, M.A.; writing—review and editing, Q.S.K. and M.J.; supervision, C.I.P.; funding acquisition, M.A. All authors have read and agreed to the published version of the manuscript.

Funding: The authors are grateful to the Raytheon Chair for Systems Engineering for funding this project.

Institutional Review Board Statement: Not applicable.

Informed Consent Statement: Not applicable.

Data Availability Statement: The data presented in this study are available on request from the corresponding author. The data are not publicly available due to large data sets.

Acknowledgments: The authors are grateful to the (Raytheon Chair for Systems Engineering) for funding. The authors are also thankful to University of Engineering and Technology, Peshawar and GIK Institute for providing necessary technical assistance.

Conflicts of Interest: The authors declare no conflict of interest.

References

1. Hovelaque, V.; Duvaleix-Tr'eguer, S.; Cordier, J. Effects of constrained supply and price contracts on agricultural cooperatives. *Eur. J. Oper. Res.* **2009**, *199*, 769–780. [CrossRef]
2. Javalgi, R.; Hall, K.; Cavusgil, S.T. Corporate entrepreneurship, customer-oriented selling, absorptive capacity, and international sales performance in the in-ternational b2b setting: Conceptual framework and research propositions. *Int. Bus. Rev.* **2014**, *23*, 1193–1202.
3. Glavas, C.; Pike, S.; Mathews, S. Internet-enabled tourism entrepreneurs. International entrepreneurial values elicited through repertory test and laddering analysis. *Int. J. Tour. Res.* **2014**, *16*, 44–55. [CrossRef]
4. Kumar, V.; Ramachandran, D.; Kumar, B. Influence of new-age technologies on marketing: A research agenda. *J. Bus. Res.* **2020**. [CrossRef]
5. Akram, U.; Hui, P.; Khan, M.K.; Tanveer, Y.; Mehmood, K.; Ahmad, W. How website quality affects online impulse buying. *Asia Pac. J. Mark. Logist.* **2018**, *30*, 235–256. [CrossRef]
6. Nakamoto, S. Bitcoin: A Peer-to-Peer Electronic Cash System. Technical Report, Manubot. 2019. Available online: https://git.dhimmel.com/bitcoin-whitepaper/ (accessed on 15 June 2020).
7. Yiyan, C.; Ye, L.; Cunjin, L. Electronic agriculture, blockchain and digital agricultural democratization: Origin, theory and application. *J. Clean. Prod.* **2020**, *268*, 122071.
8. Brakeville, S.; Perepa, B. Blockchain basics: Introduction to distributed ledgers. *Int. Bus. Mach.* **2016**, *6*, 23–52.
9. Busca, L.; Bertrandias, L. A framework for digital marketing research: Investigating the four cultural eras of digital marketing. *J. Interact. Mark.* **2020**, *49*, 1–19. [CrossRef]
10. Wells, J.D.; Valacich, J.S.; Hess, J.T. What signal are you sending? How website quality influences perceptions of product quality and purchase intentions. *MIS Q.* **2011**, *35*, 373–396. [CrossRef]
11. Xu, J.; Benbasat, I.; Cenfetelli, R.T. Integrating service quality with system and information quality: An empirical test in the e-service context. *MIS Q.* **2013**, *37*, 777–794. [CrossRef]
12. Gatautis, R.; Vaiciukynaite, E. Website atmosphere: Towards revisited taxonomy of website elements. *Econ. Manag.* **2013**, *18*, 537–544. [CrossRef]

13. Enginess. How Web Design Affects Customer Trust. 2020. Available online: https://www.enginess.io/insights/How-Web-Design-Affects-Customer-Trust (accessed on 24 July 2020).
14. Yip, G.; Dempster, A. Using the internet to enhance global strategy. *Eur. Manag. J.* **2005**, *23*, 1–13. [CrossRef]
15. Leonidou, C.L. An analysis of the barriers hindering small business export development. *J. Small Bus. Manag.* **2004**, *42*, 279–302. [CrossRef]
16. Dujak, D.; Sajter, D. Blockchain applications in supply chain. In *SMART Supply Network*; Springer: Cham, Switzerland, 2019; pp. 21–46.
17. Chang, Y.; Iakovou, E.; Shi, W. Blockchain in global supply chains and cross border trade: A critical synthesis of the state-of-the-art, challenges and opportunities. *International J. Prod. Res.* **2020**, *58*, 2082–2099. [CrossRef]
18. Turban, E.; King, D.; Lee, J.; Viehland, D. *Electronic Commerce: A Managerial Perspective 2002*; Prentice Hall: Upper Saddle River, NJ, USA, 2002; ISBN 139752854.
19. Krishnamurthy, S.; Singh, N. The international e-marketing framework (IEMF): Identifying the building blocks for future global e-marketing research. *Int. Mark. Rev.* **2005**, *22*, 605–610. [CrossRef]
20. Kalyanam, K.; McIntyre, S. The e-marketing mix: A contribution of the e-tailing wars. *J. Acad. Mark. Sci.* **2002**, *30*, 487–499. [CrossRef]
21. Yan, R. Cooperative advertising, pricing strategy and firm performance in the e-marketing age. *J. Acad. Mark. Sci.* **2010**, *38*, 510–519. [CrossRef]
22. Cannon, J.P.; Perreault, W.D., Jr. Buyer-seller relationships in business markets. *J. Mark. Res.* **1999**, *36*, 439–460.
23. Albrecht, C.C.; Dean, D.L.; Hansen, J.V. Marketplace and technology standards for b2b e-commerce: Progress, challenges, and the state of the art. *Inf. Manag.* **2005**, *42*, 865–875. [CrossRef]
24. Kalvenes, J.; Basu, A. Design of robust business-to-business electronic marketplaces with guaranteed privacy. *Manag. Sci.* **2006**, *52*, 1721–1736. [CrossRef]
25. Chong, W.K.; Bian, D.; Zhang, N. E-marketing services and e-marketing performance: The roles of innovation, knowledge complexity and environmental turbulence in influencing the relationship. *J. Mark. Manag.* **2016**, *32*, 149–178. [CrossRef]
26. Wong, C.W.Y.; Lai, K.H.; Cheng, T. Value of information integration to supply chain management: Roles of internal and external contingencies. *J. Manag. Inf. Syst.* **2011**, *28*, 161–200. [CrossRef]
27. Kaplan, S.; Sawhney, M. E-hubs: The new b2b marketplaces. *Harv. Bus. Rev.* **2000**, *78*, 97. [PubMed]
28. Cao, B.; Wang, X.; Zhang, W.; Song, H.; Lv, Z. A many-objective optimization model of industrial internet of things based on private blockchain. *IEEE Netw.* **2020**, *34*, 78–83. [CrossRef]
29. Song, J.; Zhong, Q.; Wang, W.; Su, C.; Tan, Z.; Liu, Y. FPDP: Flexible Privacy-preserving Data Publishing Scheme for Smart Agriculture. *IEEE Sens. J.* **2020**. [CrossRef]
30. Hu, X.; Chong, H.Y.; Wang, X. Sustainability perceptions of off-site manufacturing stakeholders in Australia. *J. Clean. Prod.* **2019**, *227*, 346–354. [CrossRef]
31. Soh, C.; Markus, M.L.; Goh, K.H. Electronic marketplaces and price transparency: Strategy, information technology, and success. *MIS Q.* **2006**, *30*, 705–723. [CrossRef]
32. Chen, H.; Chen, A.; Xu, L.; Xie, H.; Qiao, H.; Lin, Q.; Cai, K. A deep learning CNN architecture applied in smart near-infrared analysis of water pollution for agricultural irrigation resources. *Agric. Water Manag.* **2020**, *240*, 106303. [CrossRef]
33. Dadzie, K.Q.; Chelariu, C.; Winston, E. Customer service in the internet-enabled logistics supply chain: Website design antecedents and loyalty effects. *J. Bus. Logist.* **2005**, *26*, 53–78. [CrossRef]
34. Harrison, T.; Waite, K. A time-based assessment of the influences, uses and benefits of intermediary website adoption. *Inf. Manag.* **2006**, *43*, 1002–1013. [CrossRef]
35. Hu, J.; Luo, F.; Liao, G. A building an e-agriculture business integration platform with web services composition. In *2008 International Symposiums on Information Processing*; IEEE: New York, NY, USA, 2008; pp. 340–344.
36. Chen, P. Efficient tourism website design for supply chain management. In *International Conference on Information and Business Intelligence*; Springer: Berlin/Heidelberg, Germany, 2011; pp. 660–665.
37. Taheri, M.; Ebrahimi, S.; Khoshalhan, F. Game-theoretic analysis of supply chain coordination under advertising and price dependent demand. *Int. J. Ind. Eng. Comput.* **2014**, *5*, 395–406. [CrossRef]
38. Oberoi, P.; Patel, C.; Haon, C. Technology sourcing for website personalization: A supply- and demand-side perspective. In *Celebrating America's Pastimes: Baseball, Hot Dogs, Apple Pie and Marketing*; Springer: Cham, Switzerland, 2016; pp. 449–462.
39. Zhao, L.; Zhang, J.; Xie, J. Impact of demand price elasticity on advantages of cooperative advertising in a two-tier supply chain. *Int. J. Prod. Res.* **2016**, *54*, 2541–2551. [CrossRef]
40. Lin, Y.; Petway, J.R.; Anthony, J.; Mukhtar, H.; Liao, S.; Chou, C.; Ho, Y.F. Blockchain: The evolutionary next step for ict e-agriculture. *Environments* **2017**, *4*, 50. [CrossRef]
41. Cai, L.; He, X.; Dai, Y.; Zhu, K. Research on b2b2c e-commerce website design based on user experience. *J. Phys. Conf. Ser.* **2018**, *1087*, 062043. [CrossRef]
42. Kamilaris, A.; Fonts, A.; Prenafeta-Bold, F.X. The rise of blockchain technology in agriculture and food supply chains. *Trends Food Sci. Technol.* **2019**, *91*, 640–652. [CrossRef]
43. Xiao, D.; Zhou, Y.W.; Zhong, Y.; Xie, W. Optimal cooperative advertising and ordering policies for a two-echelon supply chain. *Comput. Ind. Eng.* **2019**, *127*, 511–519. [CrossRef]

44. Song, L.; Wang, X.; Merveille, N. Research on blockchain for sustainable e-agriculture. In *2020 IEEE Technology & Engineering Management Conference (TEMSCON)*; IEEE: New York, NY, USA, 2020; pp. 1–5.
45. Choi, T.M. Mobile-app-online-website dual channel strategies: Privacy concerns, e-payment convenience, channel relationship, and coordination. *IEEE Trans. Syst. Man Cybern. Syst.* **2020**. [CrossRef]
46. Sarkar, B.; Omair, M.; Kim, N. A cooperative advertising collaboration policy in supply chain management under uncertain conditions. *Appl. Soft Comput.* **2020**, *88*, 105948. [CrossRef]
47. Cachon, G.P.; Netessine, S. Game theory in supply chain analysis. In *Models, Methods, and Applications for Innovative Decision Making*; INFORMS Tutorials in Operations Research; INFORMS: Catonsville, MD, USA, 2006; pp. 200–233.
48. Yu, Y.; Chu, F.; Chen, H. A stackelberg game and its improvement in a VMI system with a manufacturing vendor. *Eur. J. Oper. Res.* **2009**, *192*, 929–948. [CrossRef]
49. Esmaeili, M.; Aryanezhad, M.B.; Zeephongsekul, P. A game theory approach in seller-buyer supply chain. *Eur. J. Oper. Res.* **2009**, *195*, 442–448. [CrossRef]
50. Cai, G.; Zhang, Z.G.; Zhang, M. Game theoretical perspectives on dual-channel supply chain competition with price discounts and pricing schemes. *Int. J. Prod. Econ.* **2009**, *117*, 80–96. [CrossRef]
51. Cai, G.; Chiang, W.C.; Chen, X. Game theoretic pricing and ordering decisions with partial lost sales in two-stage supply chains. *Int. J. Prod. Econ.* **2011**, *130*, 175–185. [CrossRef]
52. Banerjee, A. A joint economic-lot-size model for purchaser and vendor. *Decis. Sci.* **1986**, *17*, 292–311. [CrossRef]
53. Dada, M.; Srikanth, K.N. Pricing policies for quantity discounts. *Manag. Sci.* **1987**, *33*, 1247–1252. [CrossRef]
54. Pal, B.; Sana, S.S.; Chaudhuri, K. Three-layer supply chain—A production inventory model for reworkable items. *Appl. Math. Comput.* **2012**, *219*, 530–543. [CrossRef]
55. Kang, C.W.; Imran, M.; Omair, M.; Ahmed, W.; Ullah, M.; Sarkar, B. Stochastic-Petri Net Modeling and Optimization for Outdoor Patients in Building Sustainable Healthcare System Considering Staff Absenteeism. *Mathematics* **2019**, *7*, 499. [CrossRef]
56. Farooq, M.U.; Salman, Q.; Arshad, M.; Khan, I.; Akhtar, R.; Kim, S. An artificial bee colony algorithm based on a multi-objective framework for supplier integration. *Appl. Sci.* **2019**, *9*, 588. [CrossRef]
57. Omair, M.; Noor, S.; Hussain, I.; Maqsood, S.; Khattak, S.B.; Akhtar, R.; Haq, I.U. Sustainable development tool for Khyber Pakhtunkhwa's dimension stone industry. *Technol. J* **2015**, *20*, 160–165.
58. Bergen, M.; John, G. Understanding cooperative advertising participation rates in conventional channels. *J. Mark. Res.* **1997**, *34*, 357–369. [CrossRef]
59. Nagler, M.G. An exploratory analysis of the determinants of cooperative advertising participation rates. *Mark. Lett.* **2006**, *17*, 91–102. [CrossRef]
60. Vishal, A. Bhosale and Ravi Kant. An integrated Fuzzy Delphi and Fuzzy Inference System for ranking the solutions to overcome the supply chain knowledge flow barriers. *Int. J. Inf. Decis. Sci.* **2019**, *11*, 320–354. [CrossRef]
61. Amindoust, A.; Saghafinia, A. Textile supplier selection in sustainable supply chain using a modular fuzzy inference system model. *J. Text. Inst.* **2017**, *108*, 1250–1258. [CrossRef]
62. Paul, S.K.; Sarker, R.; Essam, D. A quantitative model for disruption mitigation in a supply chain. *Eur. J. Oper. Res.* **2017**, *257*, 881–895. [CrossRef]
63. Pourjavad, E.; Shahin, A. The application of mamdani fuzzy inference system in evaluating green supply chain management performance. *Int. J. Fuzzy Syst.* **2018**, *20*, 901–912. [CrossRef]
64. Saaty, T.L. Decision making with the analytic hierarchy process. *Int. J. Serv. Sci.* **2008**, *1*, 83–98. [CrossRef]
65. Demangeot, C.; Broderick, A.J. Engaging customers during a website visit: A model of website customer engagement. *Int. J. Retail Distrib. Manag.* **2016**, *44*, 814–839. [CrossRef]
66. Tan, F.B.; Tung, L.L.; Xu, Y. A study of web-designers' criteria for effective business-to consumer (b2c) websites using the repertory grid technique. *J. Electron. Commer. Res.* **2009**, *10*, 3–23.
67. Zhu, X.; Tan, Z. SEO keyword analysis and its application in website editing system. In *2012 8th International Conference on Wireless Communications, Networking and Mobile Computing*; IEEE: New York, NY, USA, 2012; pp. 1–4.
68. Yalçın, N.; Köse, U. What is search engine optimization: SEO? *Procedia—Soc. Behav. Sci.* **2010**, *9*, 487–493. [CrossRef]
69. Cao, M.; Zhang, Q.; Seydel, J. B2c e-commerce web site quality: An empirical examination. *Ind. Manag. Data Syst.* **2005**, *105*, 645–661. [CrossRef]
70. Miranda, F.J.; Rubio, S.; Chamorro, A. The web as a marketing tool in the Spanish foodservice industry: Evaluating the websites of Spain's top restaurants. *J. Foodserv. Bus. Res.* **2015**, *18*, 146–162. [CrossRef]
71. Wang, Y.D.; Emurian, H.H. Trust in e-commerce: Consideration of interface design factors. *J. Electron. Commer. Organ. JECO* **2005**, *3*, 42–60. [CrossRef]
72. Lowry, P.B.; Vance, A.; Moody, G.; Beckman, B.; Read, A. Explaining and predicting the impact of branding alliances and web site quality on initial consumer trust of e-commerce web sites. *J. Manag. Inf. Syst.* **2008**, *24*, 199–224. [CrossRef]
73. Shergill, G.S.; Chen, Z. Web-based shopping: Consumers' attitudes towards online shopping in New Zealand. *J. Electron. Commer. Res.* **2005**, *6*, 78.
74. Esfahani, M.M.S.; Biazaran, M.; Gharakhani, M. A game theoretic approach to coordinate pricing and vertical co-op advertising in manufacturer-retailer supply chains. *Eur. J. Oper. Res.* **2011**, *211*, 263–273.
75. Bellman, R.E.; Zadeh, L.A. Decision-making in a fuzzy environment. *Manag. Sci.* **1970**, *17*, B-141–B-273. [CrossRef]

76. Jain, R. Decision making in the presence of fuzzy variables. *IEEE Trans. Syst. Man Cybern.* **1976**, *6*, 698–703.
77. Yao, J.S.; Chiang, J. Inventory without backorder with fuzzy total cost and fuzzy storing cost defuzzified by centroid and signed distance. *Eur. J. Oper. Res.* **2003**, *148*, 401–409. [CrossRef]
78. Chang, H.C. An application of fuzzy sets theory to the EOQ model with imperfect quality items. *Comput. Oper. Res.* **2004**, *31*, 2079–2092. [CrossRef]
79. Ben-Daya, M.; As'ad, R.; Seliaman, M. An integrated production inventory model with raw material replenishment considerations in a three-layer supply chain. *Int. J. Prod. Econ.* **2013**, *143*, 53–61. [CrossRef]
80. Giannoccaro, I.; Pontrandolfo, P.; Scozzi, B. A fuzzy echelon approach for inventory management in supply chains. *Eur. J. Oper. Res.* **2003**, *149*, 185–196. [CrossRef]
81. Sarkar, B.; Ganguly, B.; Sarkar, M.; Pareek, S. Effect of variable transportation and carbon emission in a three-echelon supply chain model. *Transp. Res. Part E: Logist. Transp. Rev.* **2016**, *91*, 112–128. [CrossRef]
82. Birgin, E.G.; Haeser, G.; Ramos, A. Augmented lagrangians with constrained subproblems and convergence to second-order stationary points. *Optim. Online* **2016**, *69*, 51–75. [CrossRef]
83. Omair, M.; Ullah, M.; Ganguly, B.; Noor, S.; Maqsood, S.; Sarkar, B. The quantitative analysis of workers' stress due to working environment in the production system of the automobile part manufacturing industry. *Mathematics* **2019**, *7*, 627. [CrossRef]
84. Schittkowski, K. NLPQL: A fortran subroutine solving constrained nonlinear programming problems. *Ann. Oper. Res.* **1986**, *5*, 485–500. [CrossRef]
85. Mostafa, N.B.; Khajavi, M.N. Optimization of welding parameters for weld penetration in fcaw. *J. Achiev. Mater. Manuf. Eng.* **2006**, *16*, 132–138.
86. Theodorakatos, N.P.; Manousakis, N.M.; Korres, G.N. A sequential quadratic programming method for contingency-constrained phasor measurement unit placement. *Int. Trans. Electr. Energy Syst.* **2015**, *25*, 3185–3211. [CrossRef]
87. Sarkar, B.; Omair, M.; Choi, S.-B. A Multi-Objective Optimization of Energy, Economic, and Carbon Emission in a Production Model under Sustainable Supply Chain Management. *Appl. Sci.* **2018**, *8*, 1744. [CrossRef]
88. Omair, M.; Sarkar, B.; Cárdenas-Barrón, L.E. Minimum Quantity Lubrication and Carbon Footprint: A Step towards Sustainability. *Sustainability* **2017**, *9*, 714. [CrossRef]
89. Alkahtani, M.; Omair, M.; Khalid, Q.S.; Hussain, G.; Sarkar, B. An Agricultural Products Supply Chain Management to Optimize Resources and Carbon Emission Considering Variable Production Rate: Case of Nonperishable Corps. *Processes* **2020**, *8*, 1505. [CrossRef]
90. Ganguly, B.; Sarkar, B.; Sarkar, M.; Pareek, S.; Omair, M. Influence of controllable lead time, premium price, and unequal shipments under environmental effects in a supply chain management. *RAIRO-Oper. Res.* **2019**, *53*, 1427–1451. [CrossRef]

sustainability

Systematic Review

On Deploying Blockchain Technologies in Supply Chain Strategies and the COVID-19 Pandemic: A Systematic Literature Review and Research Outlook

Mohammad Nabipour [1,*] and M. Ali Ülkü [2]

1 Department of Civil Engineering, University of Shomal, Amol 46161-84596, Iran
2 Centre for Research in Sustainable Supply Chain Analytics, Rowe School of Business, Dalhousie University, Halifax, NS B3H 4R2, Canada; ulku@dal.ca
* Correspondence: mhd.nabipour@shomal.ac.ir; Tel.: +98-9111256197

Abstract: The emergence of a new pandemic, known as COVID-19, has touched various sections of the supply chain (SC). Since then, numerous studies have been conducted on the issue, but the need for a holistic review study that highlights the gaps and limits of previous research, as well as opportunities and agendas for future studies, is palpable. Through a systematic literature review on blockchain technology (BCT) deployment in supply-chain management (SCM) concerning the COVID-19 pandemic, this research seeks to add to the content of previous studies and to enlighten the path for future studies. Relevant papers were found using a variety of resources (Scopus, Google Scholar, Web of Science, and ProQuest). Seventy-two articles were systematically selected, considering the PRISMA procedure, and were thoroughly analyzed based on BCT, methodologies, industrial sectors, geographical, and sustainability context. According to our findings, there is a significant lack of empirical and quantitative methodologies in the literature. The majority of studies did not take specific industries into account. Furthermore, the articles focusing on the sustainability context are few, particularly regarding social and environmental issues. In addition, most of the reviewed papers did not consider the geographical context. The results indicate that the deployment of BCT in several sectors is not uniform, and this utilization is reliant on their services during the COVID-19 pandemic. Furthermore, the concentration of research on the impacts of the BCT on SCM differs according to the conditions of various countries in terms of the consequences of the COVID-19 pandemic. The findings also show that there is a direct relationship between the deployment of BCT and sustainability factors, such as economic and waste issues, under the circumstances surrounding COVID-19. Finally, this study offers research opportunities and agendas to help academics and other stakeholders to gain a better knowledge of the present literature, recognize aspects that necessitate more exploration, and drive prospective studies.

Keywords: blockchain; COVID-19 pandemic; digitalization; visibility; transparency; smart contracts; sustainability; supply chain management; literature review

Citation: Nabipour, M.; Ülkü, M.A. On Deploying Blockchain Technologies in Supply Chain Strategies and the COVID-19 Pandemic: A Systematic Literature Review and Research Outlook. *Sustainability* **2021**, *13*, 10566. https://doi.org/10.3390/su131910566

Academic Editors: Kamalakanta Muduli, Rakesh Raut, Balkrishna Eknath Narkhede and Himanshu Shee

Received: 30 July 2021
Accepted: 13 September 2021
Published: 23 September 2021

Publisher's Note: MDPI stays neutral with regard to jurisdictional claims in published maps and institutional affiliations.

Copyright: © 2021 by the authors. Licensee MDPI, Basel, Switzerland. This article is an open access article distributed under the terms and conditions of the Creative Commons Attribution (CC BY) license (https://creativecommons.org/licenses/by/4.0/).

1. Introduction

Many serious disease outbreaks have occurred in supply chains in the past; the World Health Organization (WHO) has recorded several epidemics and pandemics in the last decade alone [1]. Major public health crises can have serious consequences for enterprises and supply chains, such as lowering efficacy and productivity [2–4], and spreading ripple effects, compromising their resilience, sustainability, and long-term viability [5]. Unlike past outbreaks, COVID-19 has influenced all elements of the supply chain (SC) network at the same time, causing significant disruption in the supply chain's flow [6,7]. All sectors, including supplies, transportation, and production, have been adversely impacted by interruptions caused by quarantines, the shutting-down of various sectors, and uncertainty in supply and demand due to government restrictions. Some industries, such as energy sectors and passenger transportation, witnessed significant drops in demand [8–11]. However, the demand for essential products like healthcare materials and equipment, as well as food, has grown [12–14]. The majority of technology companies have seen a similar uptick in demand [15], due to the pandemic compelling academic organizations to utilize digital services for online teaching and forcing corporations to enable their staff to work remotely [16].

Many firms' failures to respond promptly to increasing demand because of the COVID-19 pandemic are attributable to a lack of supply chain management (SCM) capability and coordination [17]. Some studies indicate that technology has become a significant element in deciding a company's success or failure during COVID-19, implying that utilizing artificial intelligence (AI) and blockchain technologies (BCT) might assist SCs in becoming more resilient [14,17–22] and that these technologies assist businesses in preventing the negative consequences of future pandemics and preparing for unanticipated interruptions [23].

Considering the devastating effects of the COVID-19 pandemic on SCs, academics are increasingly focusing on the issue. In response to the rising interest, describing the present status of the literature and outlining future study prospects at this preliminary point might shed light on the gaps and shortcomings of previous studies and prevent future research in this subject from being duplicated [24]. A preliminary review of the research literature reveals that, despite the abundance of studies in the field of technology and the COVID-19 pandemic, there is a need for studies that can present the findings and analyze the shortcomings of previous studies in terms of blockchain technologies (BCT), research methodologies, industry sectors, geographical context and stability issues. Accordingly, the motivation of this study is to not only present an integrated picture of previous studies' findings based on various factors but also to closely examine the shortcomings of existing publications and propose research agendas that could advance the current domain of research in developing a deeper understanding and controlling the repercussions of the COVID-19 pandemic.

A systematic literature review (SLR) may aid in understanding our situation, focusing on previous results, and planning based on current flaws [25]. Through an SLR, this study aims to address the following research questions. (1) What are the key findings of the existing literature on the COVID-19 pandemic in SCM, focusing on BCT, industrial sectors, methods, sustainability, and geography categorization? (2) What are the main gaps in the existing literature in each of the aforementioned criteria, based on the analysis of the findings of previous studies? (3) What are the opportunities and suggestions for future studies on using BCT in SCM in the light of the COVID-19 pandemic?

Based on an analysis of the research literature findings, we attempted to evaluate the following hypotheses.

Hypothesis 1 (H1). *BCT deployment in different industrial sectors is not uniform; adopting one type of technology means depending on its services for that particular sector, especially during the COVID-19 pandemic.*

Hypothesis 2 (H2). *The density of research on the impacts of the BCT on SCM, taking into consideration geographical contexts, varies according to the circumstances of different nations with regard to the COVID-19 pandemic's repercussions.*

Hypothesis 3 (H3). *The deployment of BCT is linked to supply-chain sustainability under the uncertain conditions caused by the COVID-19 pandemic.*

The remainder of this work is laid out as follows. A background on BCT in SCM is presented in Section 2. The review methodology is presented in Section 3. The findings of previous studies are analyzed in Section 4. Section 5 discusses additional study prospects and focuses on the outcomes. Section 6 offers the research agendas. Finally, Section 7 concludes the paper and provides suggestions for future research.

2. The Blockchain in Supply Chain Management

Blockchain (BC) is a decentralized, distributed, and revolutionary technology that ensures the secrecy, reliability, and accessibility of all data and transactions [26]. The BCT employs a real-time cloud storage architecture [27] that allows transactions to be completed in minutes via digital platforms, all without the need for third-party verification [28]. As a peer-to-peer (P2P) distributed data infrastructure [29], BC has the potential to generate decentralized currencies, digitally automated contracts (smart contracts), and smart properties that could be managed through online services [30,31]. When the records have been added, they cannot be changed without affecting the preceding data, making them extremely secure for enterprise activities [32]. BC offers an enormous opportunity to alter every aspect of the SC, from the acquisition of raw materials through to the eventual delivery to customers [33]. It also assists in the enhancement of the effectiveness and reliability of existing digital platforms [34,35], such as the Internet of Things (IoT) and other Industry 4.0 technologies. By integrating the BC with other technologies, it is possible to establish permanent and shareable records of a product's movement through the SC processes, as well as to improve product traceability, authenticity, and legality [36].

Although the development and use of BCT are still in their infancy [37,38], the increasing volume of literature indicates that four areas of BCT are commonly used in SCs: digitalization, visibility, transparency, and smart contracts. Large numbers of supply-chain parties are involved in international commerce, and most operations are paper-based, generating delays and disrupting the effective flow of products. BCT might help digitize paper-based documents and provide shared, immutable, and real-time tracking of all transactions across network members [39]. Furthermore, by avoiding mediators, BCT may decrease verification and transaction costs [40]; the P2P network is especially beneficial for transitory commercial partnerships since it lowers the cost of developing trust [41]. The use of visibility in supply chains has been likened to real-time tracking [42,43] and timestamping, which is the act of assigning a time-related command to a series of actions [44]. Intricacy among supply-chain players is sometimes caused by geographically distant operations and trade partners. As a result, obtaining and keeping trustworthy data is important. The BC's purpose in this setting is to enable seamless networks and perfect visibility [45,46]. This reliability might have a direct impact on the safety of critical industries, like food and pharmaceutical companies [47]. Improved transparency might come in the form of cybercrime prevention and data-sharing security [47,48]. Due to the extremely delicate financial data of business transactions, transparency is essential for supply-chain participants as it may increase confidentiality, auditability, and operations performance [49,50]. Parties can secure product characterization and critical details with a transparent supply chain and restrict regulatory accessibility [51]. A smart contract is a digitalized transaction mechanism that implements the provisions of a contract automatically [41]. Whenever a cargo is received by the customer, a smart contract may deliver a payment to the supplier [52]. Smart contracts might cut down on the number of mediators and require fewer physical operations, lowering operating expenses [53]. Furthermore, smart contracts can automate operations, such as delivering copyright papers to suitable partners, giving agreed-upon contracts to designated parties for digital execution, and upgrading programs as mutually agreed regarding revisions or compensation situations [41,54].

3. Review Methodology

An SLR has been proven as a thorough methodology for review studies [25,55,56], and, by using this approach, we attempt to analyze the associated literature on the blockchain's deployment in SCM during the COVID-19 pandemic. By considering the PRISMA procedure, Figure 1 illustrates the search methodology for this study.

Relevant papers were found using a variety of resources (Scopus, Google Scholar, Web of Science, and ProQuest). We looked at research articles, literature review articles, discussion papers, opinion papers, and editorial pieces, which were available in scholarly journals. Furthermore, a reference check was performed to enhance the review resources in the final selected papers.

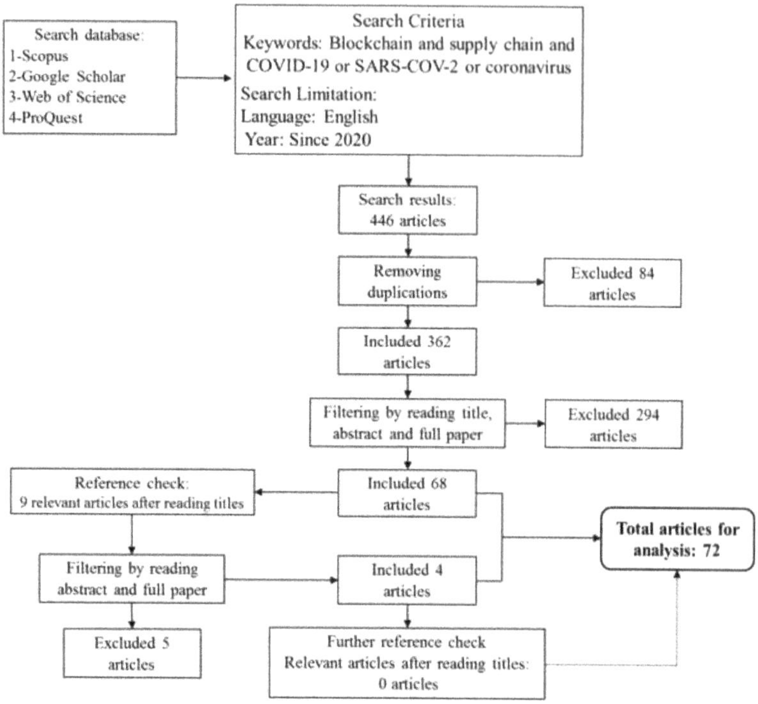

Figure 1. Adopted search methodology.

As shown in Figure 1, all selected databases were searched with the phrases "Blockchain" and "supply chain" and "COVID-19" or "SARS-CoV-2" or "coronavirus." We limited our findings to articles available exclusively in scholarly journals, published in English since 2020, with a June 2021 cut-off date. There were 446 results in the search from all databases. The irrelevant results were then eliminated. In the first phase, 84 duplicated papers were excluded. In the second step, by reviewing titles, abstracts, and complete papers, we excluded 294 more articles. This process resulted in 68 distinct articles that incorporated all the search keywords found in the body content and that also had an emphasis on the SC in light of the COVID-19 pandemic and BCT deployment. In the next phase, by checking the references of the final identified articles, we found other nine papers. Finally, from the above-mentioned procedure of filtering these extra articles, four more papers resulted. In the end, the whole procedure resulted in a total of 72 unique articles for further evaluation.

These 72 articles were fully analyzed to identify the reviewed literature's achievements in each criterion, as well as identifying the prior research's shortcomings. Selected articles were examined from four perspectives: visibility, digitalization, transparency, and smart contracts. Based on the methodology, we determined how many of the various approaches, such as empirical, qualitative, quantitative, and decision-making approaches, have been utilized in the context of these studies. The articles are divided into subsections depending on industry sectors, such as food and agriculture, healthcare, retail, construction, etc. The selected papers were analyzed based on their geographical context in order to assess the concentration of literature in different parts of the world. Finally, based on the sustainability criteria, the publications were examined based on environmental challenges, waste management, economic concerns, and social consequences to assess the degree of involvement of the research literature with these issues. In the next step, by considering the results of the analyses, the shortcomings and gaps of the existing literature were identified and discussed. Eventually, by recognizing these flaws, the future research agendas resulted in opportunities and suggestions for future studies. Figure 2 depicts the structure for this systematic review paper's analytical method.

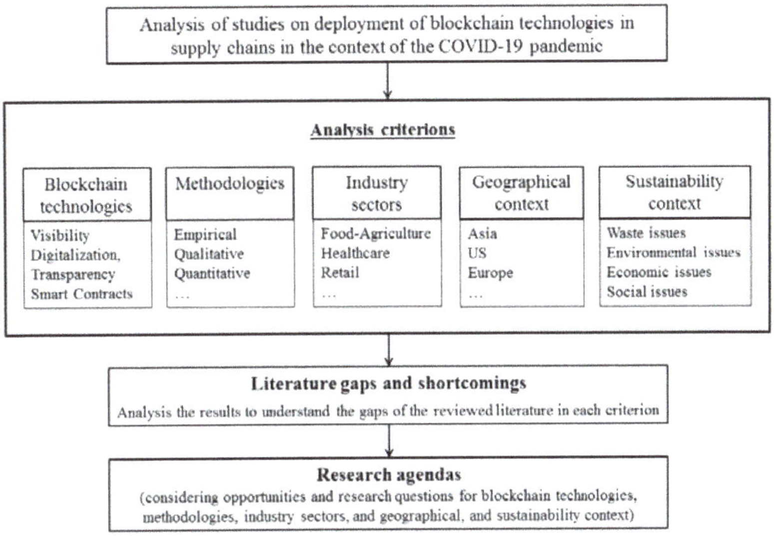

Figure 2. Structure of the analysis process.

4. Results of the Literature Review Analysis

This section examined the BCT categories, methodologies, industrial sectors, geographical contexts, and sustainability issues employed in the 72 publications identified by our searching techniques. The findings revealed those journals that have published a great deal of research on the COVID-19 pandemic in SC domains, as well as the BC topics. Table 1 illustrates the distribution of recognized articles by different references for the period from January 2020 (early stages of the pandemic) to June 2021 (completion of this research).

The contributing factors in the analysis of the selected articles are depicted in Figure 3. These factors may be found in clusters regarding BCT, methodologies, industrial sectors, geographical, and sustainable contexts, which are illustrated in different colors. There is a direct relationship between the number of occurrences where a factor appears in the reviewed papers and the size of its representative node. The greater a factor's representative node is, the more frequently it occurs in the research literature. As a result, the factors of digitalization, visibility, transparency, explanatory, literature review, food-agriculture, healthcare, and economic issues appeared most frequently in the selected articles. Figure 3 also demonstrates the connection between the investigated factors and the rate of their co-occurrence in various studies. The thicker the line demonstrating this connection, the greater the level of co-occurrence between the factors. It can be observed that the co-occurrence of digitalization, transparency and visibility in the BCT cluster is more frequent than other factors. Furthermore, there is a significant level of co-occurrence between these three factors and the explanatory and literature review factors in the methodology cluster, the food-agriculture and healthcare factors in the Industrial Sectors cluster, and the economic issues factor in the sustainability cluster. Appendix A contains descriptive analyses of the identified papers, including a summary of the articles' goals and findings (Table A1), the publishing characteristics, such as the BCT categories, methods, industries, geographical context, and sustainability issues (Table A2).

Table 1. Selected publications based on the journal's title.

Journal Title	Number of Documents
International Journal of Production Economics	5
IEEE Engineering Management Review	4
Transportation Research Part E: Logistics and Transportation Review	4
Sustainability	3
IEEE Access	3
International Journal of Production Research	2
International Journal of Information Management	2
International Journal of Environmental Research and Public Health	2
International Journal of Physical Distribution and Logistics Management	2
Journal of Cleaner Production	2
Journal of Enterprise Information Management	2
Journal of Supply Chain Management	2
Sustainable Production and Consumption	2
International Journal of Logistics Research and Applications	2
International Journal of Operations and Production Management	2
Business Strategy and the Environment	1
Industrial Management and Data Systems	1
Applied Sciences	1
Big Data and Cognitive Computing	1
Continuity and Resilience Review	1
Critical Perspectives on International Business	1
Current Research in Food Science	1
Designs	1
Diabetes and Metabolic Syndrome: Clinical Research and Reviews	1
Environment Systems and Decisions	1
Environmental Science and Pollution Research	1
European Journal of Business and Management	1
Foods	1
Frontiers of Business Research in China	1
Future Internet	1
Information	1
International Journal of Integrated Supply Management	1
International Journal of Productivity and Performance Management	1
JMIR Public Health and Surveillance	1
Journal of Business Research	1
Journal of Food Quality	1
Journal of Humanitarian Logistics and Supply Chain Management	1
Journal of Property Investment and Finance	1
Journal of Purchasing and Supply Management	1
Journal of Scientific Research and Reports	1
Journal of Supply Chain Management Systems	1
Manufacturing and Service Operations Management	1
Modern Supply Chain Research and Applications	1
Production and Operations Management	1
Production Planning and Control	1
The International Journal of Logistics Management	1
Annals of Operations Research	1
Science of the Total Environment	1
Total	72

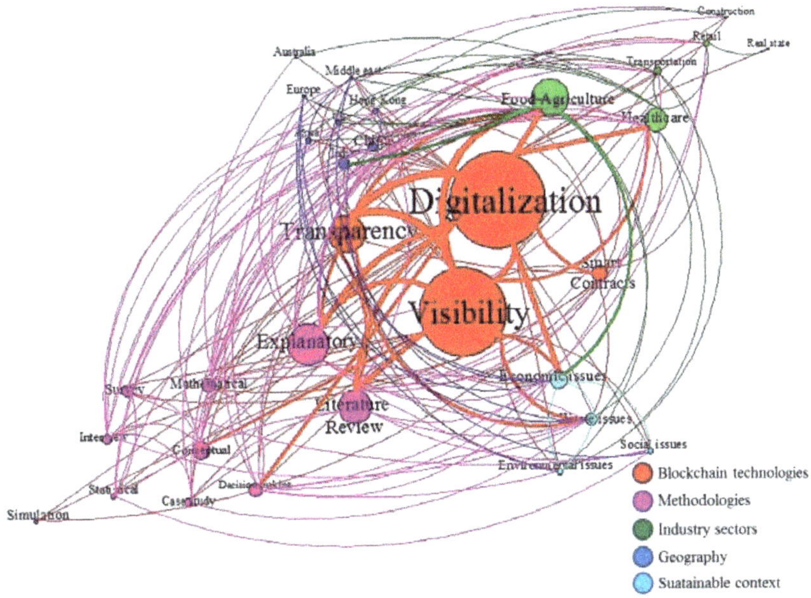

Figure 3. The co-occurrence of the contributing factors in the analysis of the selected articles.

4.1. Blockchain Technologies

In this study, we assessed the reviewed publications from the standpoint of BCT deployment. Articles are grouped into categories based on BCT, such as visibility, digitalization, transparency, and smart contracts. The results are demonstrated in Figure 4. Several studies (almost 52%) have linked various technologies, and these articles are categorized as multiple technologies (e.g., digitalization and visibility, digitalization and smart contracts, visibility and transparency, digitalization, visibility and transparency, digitalization, visibility and smart contracts). We split this category into different BCT combinations for further analysis. Table 2 demonstrates the findings of the reviewed articles on BCT deployment.

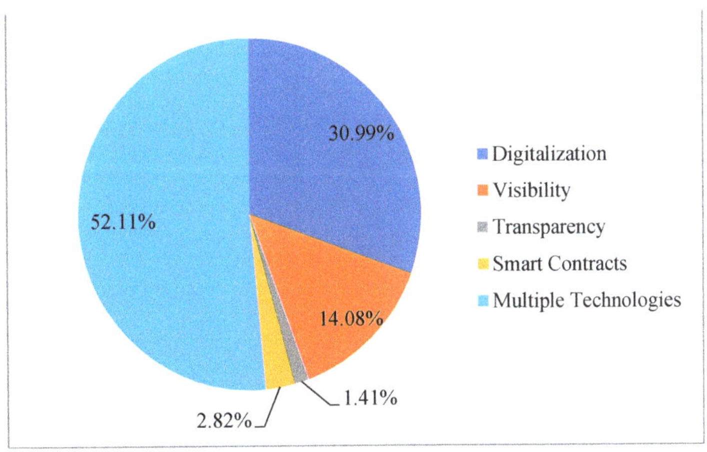

Figure 4. The proportion of reviewed articles on the subject of BCT.

The findings indicate that, of all the BCT categories, digitalization had the highest frequency in the reviewed studies, with 22 articles out of 72 [3,13,57–76]. With ten articles, visibility was also one of the most common BCT [5,20,77–84]. Sixteen of the papers looked at the impact of both digitization and visibility on the SC and the COVID-19 pandemic simultaneously [5,23,85–98]. While transparency technology has received little attention in prior research, and only one study has addressed this issue [99], the combination of visibility and transparency has been the subject of nine studies in the literature that was reviewed [98,100–107]. Six studies examined the implementation of digitalization, visibility and transparency technologies in SC and the critical circumstances of COVID-19 [18,22,108–111].

Table 2. BCT used in the reviewed studies.

Blockchain Technologies	Number of Documents	References
Digitalization	22	Bahn et al. [57]; Betcheva et al. [58]; Butt [59]; Chamola et al. [13]; Chowdhury et al. [60]; Cordeiro et al. [61]; Fusco et al. [62]; Guo et al. [63]; Karmaker et al. [64]; Kumar [65]; Kumar and Kumar Singh [66]; Labaran and Hamma-Adama [67]; Liu et al. [68]; Narayanamurthy and Tortorella [69]; Papadopoulos et al. [70]; Paul and Chowdhury [3]; Quayson et al. [71]; Sharma et al. [72]; Siriwardhana et al. [73]; Starr et al. [74]; Sufian et al. [75]; Yeganeh [76]
Visibility	10	Acioli et al. [77]; Agarwal et al. [78]; Akhigbe et al. [79]; Finkenstadt and Handfield [80]; Golan et al. [20]; Ivanov [7]; Kovács and Falagara Sigala [81]; Lin et al. [82]; Memon et al. [83]; Yang et al. [84]
Transparency	1	Kumar et al. [98]
Smart contracts	2	Ahmad et al. [111]; Lohmer et al. [21]
Digitalization and visibility	16	Choi [99]; de Sousa Jabbour et al. [23]; Di Vaio et al. [85]; Dutta et al. [86]; Gurbuz and Ozkan [87]; Ivanov and Das [88]; Ivanov and Dolgui [5]; Kazancoglu et al. [89]; Kumar and Pundir [90]; Nandi et al. [91]; Nandi et al. [92]; Platt et al. [93]; Tasnim [94]; Xu et al. [95]; Yousif et al. [96]; Zouari et al. [97]
Digitalization and smart contracts	1	Kalla et al. [112]
Visibility and transparency	9	Bakalis et al. [99]; Bumblauskas et al. [100]; Choi [27]; Choi [101]; Iftekhar et al. [102]; Kemp et al. [103]; Montecchi et al. [104]; Pillai and Mohan [105]; Sodhi and Tang [106]
Visibility and smart contracts	1	Lin et al. [113]
Digitalization, visibility and transparency	6	Etemadi et al. [107]; Ivanov and Dolgui [18]; Remko [22]; Rowan and Galanakis [108]; Yadav et al. [109]; Zhang et al. [110]
Digitalization, visibility and smart contracts	1	Bekrar et al. [114]
Digitalization, transparency and smart contracts	1	Dolgui and Ivanov [115]
All blockchain technologies	2	Bamakan et al. [116]; Sharma et al. [14]

Smart-contact technology, on the other hand, is one of the areas of BC technology that has received some slight attention in the literature, resulting in only two studies [21,112]. Furthermore, one article examining digitalization and smart contracts [113], another article evaluating visibility and smart contracts [114], one article assessing digitalization, visibility, and smart contact technologies [115], and one research topic also looked into digitalization, transparency, and smart contract technologies [116]. Finally, two of the reviewed articles dealt with all the BCTs at the same time [14,117].

4.2. Methodologies

By reviewing the past literature in terms of research methodologies, this study attempts to evaluate the contribution of each of the approaches employed in connection to the BCT in SC and to the COVID-19 pandemic. These findings may be used to discover which techniques have been used more frequently and which methods deserve more attention. Furthermore, the data can serve as a reference for many researchers in picking the appropriate methodologies. Reviewed article methods were classified into several categories, such as empirical, quantitative, qualitative, decision making, review, and multiple methods. Figure 5 demonstrates the application rates of each of these approaches, with qualitative and review methodologies having the highest share in prior studies. After that, quantitative approaches and the use of multiple methodologies were the runners-up. The findings of the methodological evaluation of the reviewed publications are shown in Table 3.

Table 3. Summary of the methodologies employed in the reviewed studies.

Methodology	Particular Field	Number of Documents	References
Empirical	Case study	2	Betcheva et al. [58]; Bumblauskas et al. [100]
	Survey	3	Guo et al. [63]; Lin et al. [113]; Yang et al. [84]
Quantitative	Simulation	2	Ivanov and Das [88]; Lohmer et al. [21]
	Mathematical	6	Ahmad et al. [111]; Choi [27]; Ivanov and Dolgui [116]; Karmaker et al. [64]; Kumar [65]; Paul and Chowdhury [3]
Qualitative	Explanatory	20	Bahn et al. [57]; Butt [59]; Choi [27]; Choi [101]; Dolgui and Ivanov [115]; Kalla et al. [112]; Kovács and Falagara Sigala [81]; Liu et al. [68]; Nandi et al. [91]; Nandi et al. [92]; Papadopoulos et al. [70]; Pillai and Mohan [105]; Platt et al. [93]; Quayson et al. [71]; Rowan and Galanakis [108]; Sharma et al. [14]; Siriwardhana et al. [73]; Sodhi and Tang [106]; Tasnim [94]; Zhang et al. [110]
	Conceptual	6	de Sousa Jabbour et al. [23]; Finkenstadt and Handfield [80]; Iftekhar et al. [102]; Ivanov [7]; Ivanov and Dolgui [116]; Starr et al. [74]
	Interview	3	Labaran and Hamma-Adama [67]; Remko [22]; Yeganeh [76]
Decision-making	MCDM Methods	5	Agarwal et al. [78]; Kazancoglu et al. [89]; Kumar and Kumar Singh [66]; Yadav et al. [109]; Zouari et al. [97]
	SWOT	1	Fusco et al. [62]
Review	Literature Review/Systematic Literature Review	15	Acioli et al. [77]; Akhigbe et al. [79]; Bakalis et al. [99]; Bekrar et al. [114]; Chamola et al. [13]; Chowdhury et al. [60]; Cordeiro et al. [61]; Di Vaio et al. [85]; Dutta et al. [86]; Etemadi et al. [107]; Golan et al. [20]; Kumar et al. [98]; Kumar and Pundir [90]; Montecchi et al. [104]; Yousif et al. [96]
Multiple methods	Explanatory and Statistical	1	Memon et al. [83]
	Conceptual and Survey	1	Lin et al. [82]
	Conceptual and Interview	1	Xu et al. [95]
	Conceptual and Review	1	Sufian et al. [75]
	Case study and Review	2	Bamakan et al. [116]; Kemp et al. [103]
	Mathematical and Decision-making	1	Sharma et al. [72]
	Survey and Statistical	2	Gurbuz and Ozkan [87]; Narayanamurthy and Tortorella [69]

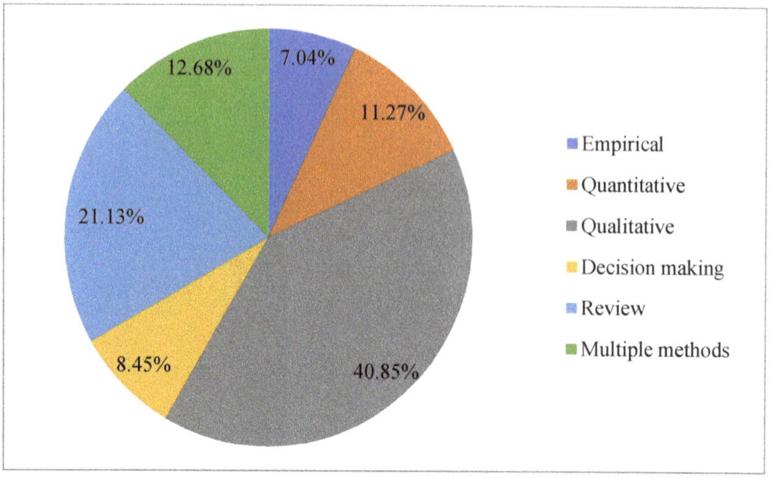

Figure 5. The proportion of reviewed articles based on the research methodologies.

4.3. *Industry Sectors*

Examining the relation between BCT and various industries may aid in understanding the significance of these technologies, particularly in light of the COVID-19 outbreak. According to our findings, the food and agricultural sector has the most commonly utilized BCT, with 19 articles addressing COVID-19 conditions [5,56,57,66,71,79,82,83,85,87,89,94,100,101,103,106,109,110,114]. With 12 research papers since the outbreak, the healthcare industry has made a significant contribution to the reviewed articles. Eight of them were solely on healthcare matters [3,58,62,67,80,93,112,117], while two dealt with healthcare and transportation [81,96] and one dealt with the healthcare and retail sector [99]. Eight articles discuss various areas of the industry. Studies that covered more than two sectors at the same time were included in the multiple industries category [13,20,60,61,63,84,86,91]. From the findings, 26 of the 72 reviewed publications did not identify a specific sector and instead discussed the general function of BCT in SC and its applications in the management of COVID-19 situations. Table 4 summarizes the industrial sectors covered in the examined papers.

Table 4. Summary of industries discussed in the examined studies.

Industrial Sector	Number of Documents	References
Food–Agriculture	19	Akhigbe et al. [79]; Bahn et al. [57]; Bakalis et al. [99]; Bumblauskas et al. [100]; Di Vaio et al. [85]; Gurbuz and Ozkan [87]; Iftekhar et al. [102]; Ivanov and Dolgui [116]; Kazancoglu et al. [89]; Kumar [65]; Kumar and Kumar Singh [66]; Lin et al. [82]; Lin et al. [113]; Memon et al. [83]; Pillai and Mohan [105]; Quayson et al. [71]; Rowan and Galanakis [108]; Tasnim [94]; Yadav et al. [109]
Healthcare	8	Ahmad et al. [111]; Bamakan et al. [116]; Betcheva et al. [58]; Finkenstadt and Handfield [80]; Fusco et al. [62]; Labaran and Hamma-Adama [67]; Paul and Chowdhury [3]; Platt et al. [93]
Healthcare and Transportation	2	Kovács and Falagara Sigala [81]; Yousif et al. [96]
Healthcare and Retail	1	Kumar et al. [98]
Retail	1	Sharma et al. [72]
Retail and Real estate	1	Starr et al. [74]

Table 4. *Cont.*

Industrial Sector	Number of Documents	References
Construction	1	Xu et al. [95]
Mining	1	Kemp et al. [103]
automotive	1	Agarwal et al. [78]
Transportation	2	Bekrar et al. [114]; Choi [27]
Multiple industries	8	Chamola et al. [13]; Chowdhury et al. [60]; Cordeiro et al. [61]; Dutta et al. [86]; Golan et al. [20]; Guo et al. [63]; Nandi et al. [91]; Yang et al. [84]

4.4. Geographical Context

Geographical context is a key element for establishing tailored methods for managing the COVID-19 pandemic, considering that several nations have witnessed varied rates of infection and chosen differing lockdown measures to handle the pandemic crisis. Figure 6 shows the reviewed articles according to their geographical classification.

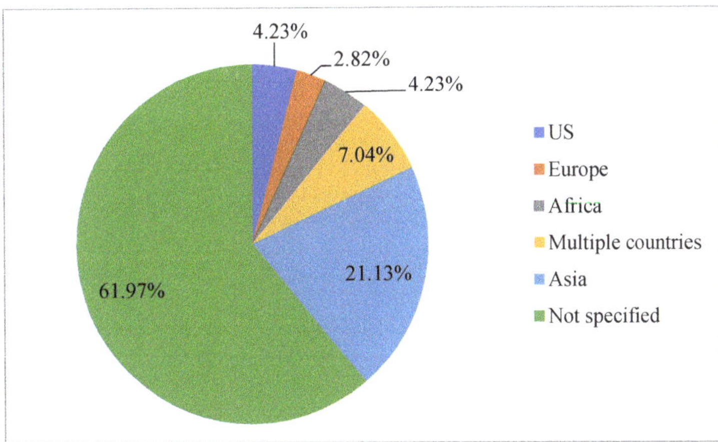

Figure 6. The proportion of reviewed articles, based on the geographical context.

According to the findings, the majority of the existing publications did not focus on the national level. As a result, 44 out of 72 papers do not provide any geographical classification in their study findings. Given the novelty of the issues covered, the scarcity of information, and the research method's heavy emphasis on qualitative and literature review, this approach is acceptable. However, some articles have looked at the outcomes on a country-by-country basis. Five papers investigated the findings in different regions of India [65,66,69,78,106]. China has been the subject of three studies [63,82,114]. The situation in Hong Kong is discussed in two articles [27,95]. The findings for these countries have been examined in certain papers, for example, one article on China and Hong Kong [84] and another on India and China [83]. Three articles were published in the United States in particular [22,80,101]. The findings in Africa and the Middle East were assessed in four papers—two in Africa [67,71], one in the Middle East [117], and one in both areas [57]. With regard to Europe, one article related to the situation in France [97] and another to Ireland [109]. Furthermore, certain articles discussed the situation in Australia [3] and Bangladesh [64]. In five studies, researchers looked at more than two nations at the same time, resulting in the multiple countries category [13,58,60,88,104]. Table 5 shows the results of the review of articles, depending on the geographical context.

Table 5. The geographical context, as discussed in the reviewed studies.

Geographical Context		Number of Documents	References
Asia	India	5	Agarwal et al. [78]; Kumar [65]; Kumar and Kumar Singh [66]; Narayanamurthy and Tortorella [69]; Pillai and Mohan [106]
	China	3	Guo et al. [63]; Lin et al. [82]; Lin et al. [113]
	Hong Kong	2	Choi [27]; Xu et al. [95]
	Middle East	1	Bamakan et al. [116]
	Bangladesh	1	Karmaker et al. [64]
	China and Hong Kong	1	Yang et al. [84]
	India and China	1	Memon et al. [83]
	Middle East and Africa	1	Bahn et al. [57]
US		3	Bumblauskas et al. [100]; Finkenstadt and Handfield [80]; Remko [22]
Europe	Ireland	1	Rowan and Galanakis [108]
	France	1	Zouari et al. [97]
Australia		1	Paul and Chowdhury [3]
Africa		2	Quayson et al. [71]; Labaran and Hamma-Adama [67]
Multiple countries		5	Betcheva et al. [58]; Chamola et al. [13]; Chowdhury et al. [60]; Ivanov and Das [88]; Kemp et al. [103]

4.5. Sustainability Context

The ability of a company to maintain self-sufficiency in commercial activities amid uncertain economic conditions, such as the COVID-19 pandemic, is dependent on the efficient control of resources engaged in SCM [118]. Regarding sustainability issues, reviewed publications are categorized with certain sustainability topics, such as actions aimed at minimizing harmful environmental consequences (such as climate change and pollution), waste management, economic concerns, and mitigating adverse social effects.

According to the findings, only 16 of 72 articles focus on the sustainability context by considering the deployment of BCT under the conditions caused by COVID-19. Five papers examined the economic concerns [23,57,64,85,92]. Four of them concentrated on waste management [7,104,112,117]. Two articles discussed social issues, such as job insecurity and employee performance [69,77]. Environmental problems were the subject of one paper [81]. Four of the articles covered more than one issue. One looked into waste and economic issues [89], another focused on economic and social subjects [76], and two investigated waste, environmental, and economic issues [109,110]. Table 6 demonstrates the results of the review articles discussing sustainability issues.

Table 6. The sustainability context, as discussed in the reviewed studies.

Sustainability Context	Number of Documents	References
Waste Issues	4	Ahmad et al. [111]; Bamakan et al. [116]; Ivanov [7]; Kemp et al. [103]
Environmental Issues	1	Kovács and Falagara Sigala, [81]
Economic Issues	5	Karmaker et al. [64]; Nandi et al. [92]; Bahn et al. [57]; de Sousa Jabbour et al. [23]; Di Vaio et al. [85]
Social Issues	2	Acioli et al. [77]; Narayanamurthy and Tortorella [69]
Waste and Economic Issues	1	Kazancoglu et al. [89]
Economic and Social Issues	1	Yeganeh et al. [76]
Waste, Environmental, and Economic Issues	2	Rowan and Galanakis [108]; Yadav et al. [109]

5. Discussion

5.1. Blockchain Technologies and Industry

We categorized BCT into four groups based on prior research findings: visibility, digitalization, transparency, and smart contracts. According to Table 2, digitalization and visibility have received the most attention in past studies. In total, 48 out of 72 articles deal either independently or jointly with these two themes. Transparency and smart contracts, on the other hand, were in a lesser proportion. One of the primary explanations for this unequal distribution might be the depth of deployment of these technologies in various industries. Analyzing the links between BCT's deployment in different industrial sectors during the COVID-19 pandemic might be a key criterion for determining the importance of these technologies in handling uncertainties in various areas of the industry. Based on the findings, the food-agricultural and healthcare sectors had the highest share of studies, with 19 and 12 publications, respectively. The results in Figure 7 illustrate the connections between BCT and different sectors of industry. Digital infrastructure, remote monitoring, and tracking are just a few of the reasons why these technologies are being given more attention. The most urgent difficulties during the COVID-19 pandemic were a lack of transparency and visibility in essential demand, production capacity, distribution constraints, and storage conditions, particularly in the food-agricultural and healthcare sectors [3,119]. As a result, the demand for real-time data, monitoring, and tracking has increased. On the other hand, the fast growth in demand for goods provided by these two industries, as well as the risk of counterfeited critical medical and pharmaceutical products, has increased the significance of these technologies and their integration with transparency [57,102]. Another reason for the increased focus on digitization was the necessity of working from home (WFH) during the COVID-19 pandemic quarantine, as well as the need for various enterprises to adapt to a digital infrastructure [69,77]. Smart contracts, which focus primarily on digital contracts and transactions, are among the technologies that have garnered the least attention from academics. Two analyses may be used for further assessment. Firstly, in terms of the impact of COVID-19, this technology provides fewer services than the other technologies described for controlling the uncertain circumstances of the supply chain, as it is primarily concerned with reducing the duration of the supply chain process [41,52]. Secondly, research on COVID-19 and BCT is still in its early phases. Consequently, as more information from various industries and their performance in facing the pandemic becomes available, smart contact technology is likely to receive a larger proportion of research than it does now.

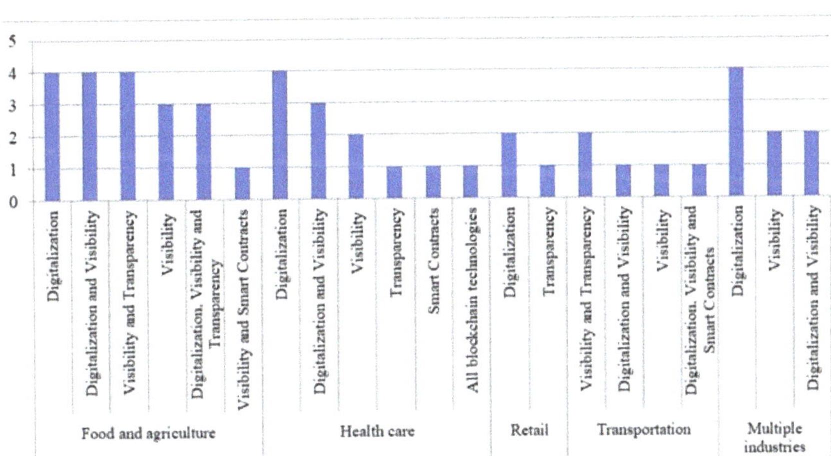

Figure 7. The links between BCT and different sectors of industry.

Given the above, it can be inferred that the first hypothesis of this study is valid and that the use of BCT is dependent on their services in various industries, particularly during COVID-19.

5.2. Methodologies and Geographical Contexts

Several categories have been suggested based on the previously reviewed methodologies (Table 3). Based on our results, the highest proportion of articles reviewed (35 out of 72) relates to the qualitative and review categories. The novel nature of both BCT and COVID-19 is part of the reason for so much attention being given to these two methodologies. Given the early stages of knowledge, qualitative assessments and the gathering of past findings might be of significant importance in advancing these issues, and the absence of using other techniques does not imply the inadequacy of present investigations. As time passes and more information is accessible, we should expect additional research based on empirical and quantitative methodologies from the application of BCT in various industries, particularly under uncertain situations like the COVID-19 pandemic.

By evaluating the studies geographically, it is clear that a considerable number of the articles assessed pertain to the Asian nations (15 out of 72), most commonly from China and India. The significant prevalence of COVID-19 and its rapid expansion, as well as their high populations and therefore high demand for vaccinations, are among the reasons why researchers are paying more attention to such countries. On the other hand, the outcomes revealed that studies in other countries have been sparse. In this regard, it should be emphasized that there is still a scarcity of data on how various nations are facing the COVID-19 outbreak. Because of this difficulty, a large number of publications used qualitative methodologies, and, as a consequence, they have not examined the effects of these technologies on a specific country's performance (44 out of 72).

Although the number of articles in the research literature that looked at geographical contexts is not large enough to definitively support the second hypothesis of this study, the findings of the reviewed articles indicate that countries like China, India, the US and the Middle East, that faced the most challenges during the COVID-19 pandemic, have received more attention from researchers than any other country. As a result, the authors believe that, based on existing evidence, the second hypothesis can be verified.

5.3. Sustainability Context

Previous studies were examined based on four sustainability criteria: waste issues, environmental challenges, economic difficulties, and social concerns. The findings revealed that only a small proportion of peer-reviewed papers addressed these topics (16 out of 72). Economic and waste issues, for example, have garnered increasing interest from academics. This is because the COVID-19 pandemic has had a direct impact on SC economics; paying attention to this issue and developing suitable measures might assist businesses to survive longer. The nature of BCT, on the other hand, is compatible with waste management, and its implementation might have a beneficial impact on SC performance. Despite the fact that the novelty of COVID-19 and its unknown impacts on various areas of the SC may have resulted in a distinct lack of study in sustainable SC, there is a definite need to focus on this area of literature.

Accordingly, we may conclude that our third hypothesis is accurate as well. Although the existing literature cannot demonstrate the impacts of BCT for all factors of sustainability criteria under the circumstances of COVID-19, it is apparent that these effects are significant in the economic and waste-management areas.

6. Research Agendas for Technology Deployment in SCM

The findings point to a plethora of potential study topics in the COVID-19 pandemic considering the deployment of BCT in SCs. Although numerous studies have been posted since the COVID-19 outbreak, additional study is still needed in several research subjects and sectors. This section aims to elucidate some of these important topics. Table 7 summarizes the most crucial future research issues and prospects in several disciplines.

Table 7. Future research agendas.

Studies Context	Research Questions and Opportunities	Suggestions
BCT context	(1) How do transparency and smart contract technologies impact the various sectors of the industry considering the COVID-19 pandemic? (2) What are the effects of BCT deployment during the COVID-19 pandemic on SCs, especially for high-demand items? (3) How do the COVID-19 circumstances affect the willingness of several sectors of industries to use BCT? (4) How might the BCT deployment will lead to a more resilient SC for the post-COVID-19 era?	Circumstances: the necessity for WFH, traceability and transparency
	(5) What are the factors that impact blockchain adoption in several sectors from the perspective of COVID-19?	Factors: BCT's services in uncertain conditions, Infrastructure availability, Cost-benefit features
Industrial contexts	(1) How does the BCT deployment affect the agricultural-food and healthcare sectors considering the COVID-19 circumstances, particularly transparency and smart contact technologies? (2) How has the retail sector been influenced by BCT usage?	Subject: Online shopping services
	(3) How do transportation and logistics engage with the usage of BCT during the COVID-19 pandemic?	Subjects: Domestic and transnational freight transportation services Uncertainty border restrictions, the uncertainty of supply and demand
	(4) How does the COVID-19 pandemic impact the technology and communication sectors' markets?	Subjects: Working from home requirements Social distancing necessity
	(5) In light of the COVID-19 pandemic, how is the energy sector, such as the oil business, engaging with the BCT?	Subjects: Border closure influences Quarantine and avoiding the use of passenger transportation systems
Methodological contexts	(1) How could practical information from different industries with regard to deploying BCT be collected at the early stages of the COVID-19 pandemic?	Methodologies: Use interview method in very early phases Use surveys in the next phases Use of decision-making methods
	(2) What are the practical findings of various industries' performances during and after the COVID-19 pandemic considering BCT deployment?	Methodologies: Use statistics methods to estimate predictions, such as regression models and data learning models Use mathematical methods to analyze choice behaviors, such as logit models and stated preference experiments
Geographical contexts	(1) What are the geographical findings of various industries' performance for BCT and COVID-19?	Subjects: Case studies from various sectors in various countries Literature reviews on the practices, and policies, and achievements of various industries in various nations, as well as the causes of their success or failure, considering BCT usage

Table 7. *Cont.*

Studies Context	Research Questions and Opportunities	Suggestions
	(2) What are the findings regarding government performance in connection to BCT during and after the COVID-19 pandemic?	Subjects: Evaluate functional solutions, and identify both effective and ineffective policies at the government level From a government viewpoint, investigating the required infrastructure and investments in the field of BCT in various businesses for the post-Covid-19 period, as well as coping with future pandemics
Sustainability context	(1) How could BCT help maintain a sustainable supply chain in the face of disruptions like the COVID-19 pandemic?	Subjects: New conceptual models and qualitative approaches, case studies, and quantitative results Policies, strategies, and governance supports in the promotion of SSC via BCT
	(2) How may the deployment of various BCTs improve the performance of sustainable supply chains during and after the COVID-19 pandemic?	Subjects: Data safety and security to improve the circular economy with the use of BCT, such as visibility and transparency Digitalized documentation for effective take-back and closed-loop supply chains using smart contracts
	(3) How may the implementation of BCT help to alleviate environmental concerns (such as climate change and GHG emissions) before, during, and after COVID-19?	Subjects: monitoring the environmental pollutions through BCT from supply chains and logistics Understanding new waste management strategies through BCT
	(4) How does the BCT influence societal concerns like job insecurity and employee performance during the COVID-19 pandemic?	Subjects: case studies and interviews with people from various industries to see how working from home and utilizing BCT affects their performance The influence of insurance plans and government assistance on job security concerns

6.1. Blockchain Technological Context

Although several studies have been conducted in various sectors of BCT, including digitization and visibility, there remain gaps, particularly in transparency and smart contracts. Previous research has demonstrated that by digitization and visualization, effective services can be offered to SCs, particularly in crisis management, such as the COVID-19 pandemic. However, these definitions are incomplete for the transparency and smart contract domains, and additional study is needed in these two areas of technology. On the other hand, the majority of studies focus on evaluating the effects of various technologies under the existing circumstances of COVID-19. However, it is expected that different industries' approaches to the deployment of other technologies would alter in the post-COVID-19 era [7,22,87]. Several experiences, such as the necessity for WFH, the benefits of traceability and transparency, and the need for confirming the authenticity of certain products, will have a significant influence on various businesses' readiness to adopt these technologies in both the private and public sectors. As a result, the present findings on these issues are unclear, and further research is needed in this field. Furthermore, previous research findings have not identified which factors might influence the deployment of these technologies in different industries under the COVID-19 pandemic (such as services in managing uncertain conditions, the existing infrastructure in companies, the relevance of these technologies with various sectors, cost-benefit studies for each of the technology categories, and so on).

6.2. Industrial Contexts

The food and health sectors have a significant share in the reviewed studies. However, when examining the use of various technologies in these two areas, it is evident that transparency and smart contact technologies have received little attention from scholars. On the other hand, other industries were greatly impacted by the deployment of technology during the COVID-19 pandemic, but this is rarely addressed in the examined papers. Because of quarantine and the necessity for social distancing, the retail sector has developed online shopping services [72,120,121]. The transportation and logistics industry is another field that was engaged in BCT during the COVID-19 pandemic, particularly those businesses that provide domestic and transnational freight transportation services [88,115]. Uncertainty regarding border restrictions, as well as diverse supply and demand circumstances, raised the necessity for these businesses to employ additional technologies [20,22]. Working from home has prompted many businesses of all kinds to turn to digitalization services, and this has had a huge influence on the technology and communication sectors' markets [69]. The energy sector, such as the oil industry, has been severely impacted by the COVID-19 pandemic. Border closures, quarantine, and the avoidance of passenger transportation systems have thrown off the balance of oil supply and demand, entirely disrupting the supply chain for this product [86,120]. Despite these significant effects, few articles concentrated on the deployment of BCT in these areas, and future studies will need to address all these issues.

6.3. Methodological Contexts

Given that COVID-19 is a fresh issue regarding all fields of knowledge, its novelty in the supply chain has also had a major impact, resulting in qualitative approaches that encompass a large proportion of the reviewed studies [22,27,57,80,102]. However, when knowledge regarding the performance of various sectors of different industries grows, it is critical for researchers to provide practical achievements regarding the topic using empirical and quantitative techniques. Furthermore, under the circumstance of a lack of practical data, methodologies such as surveys may be used to gather enough data to assemble studies using other approaches, such as statistical modeling, decision-making, and mathematical methods.

6.4. Geographical Contexts

According to our findings, researchers have not given much attention to BCT in SC under the influences of COVID-19 at the country level, because 44 publications out of 72 did not consider this issue at all. India and China, two Asian nations that have been the worst hit by the COVID-19 pandemic, have done more research on the topic than other countries [63,65,66,69,78,82,106,114]. The number of studies for developed nations such as Europe, the United States, and Australia, on the other hand, has been quite limited [22,29,97,109]. Although there is currently inadequate evidence available on industry performance in many nations, a broader examination, maybe from a governmental perspective, might lead to more comprehensive national performance studies. Some highlighted issues, such as evaluating government regulations on the usage of BCT, providing financial assistance to improve different industries infrastructures for accommodating these technologies, assessing the impacts of employing technology in the food-agricultural and healthcare sectors under the supervision of government organizations, and identifying the policies necessary in different nations for the post-COVID-19 period, might be addressed in future studies.

6.5. Sustainability Context

The reviewed articles are discussed from the perspective of certain sustainability issues, including environmental difficulties, waste management, economic concerns, and societal challenges. Although the use of BCT in supply-chain and logistics management has enhanced the sustainability of various businesses in recent years, the new conditions that have emerged in various industries, influenced by the COVID-19 pandemic, have shown the need for further research in this area. According to our findings, only a tiny percentage of the papers evaluated address the sustainability context [7,23,57,64,85,92,104]. However, when it comes to environmental and social concerns, this gap is clearly reflected [69,77,81]. This dilemma is exacerbated when we consider the societal issues that have arisen as a result of the COVID-19 pandemic. Working remotely as a result of quarantine, which is linked with BCT deployment, has intensified issues like employee performance and job insecurity. Environmental challenges, including climate change and greenhouse gas emissions, which have long been supply-chain concerns, have grown more complicated as a result of the COVID-19 pandemic. However, the application of BCT to these difficulties, particularly in current and post-pandemic situations, is ambiguous, and the study literature has paid little attention to them.

7. Conclusions and Future Works

In this paper, we investigated the state of practice for BCT deployment on SCM by focusing on the COVID-19 pandemic. In total, 72 articles were systematically selected and thoroughly analyzed, based on BCT, methodologies, industrial sectors, geographical classifications, and the sustainability context. The following is a summary of the research findings.

7.1. Theoretical and Practical Findings

We categorized BCT into four groups, including visibility, digitalization, transparency, and smart contracts. A large proportion of the reviewed articles have focused on digitalization and visibility; however, transparency and smart contracts had a smaller share. The findings reveal that this unequal distribution may be explained by the linkages between BCT deployments in various industry sectors during the COVID-19 pandemic. This finding supports our first hypothesis that BCT's utilization is reliant on the services offered in a variety of sectors, especially during COVID-19. The results show that in the food and healthcare industries, digitalization and visualization technologies have been most frequently used. This direct relationship could be due to sudden changes in supply and demand, production capacity, distribution constraints, and storage conditions, resulting in a greater demand for real-time data, monitoring, and tracking. Another reason for the growing focus on digitization and visibility is the requirement for diverse businesses to adapt to digital infrastructures, particularly when COVID-19 necessitated working from home.

We noticed that most research emphasized the overall role of BCT in SC and its applications in the control of the COVID-19 pandemic rather than a specific industry sector. We think the industry-focus approach will become more important as it is very likely that consumption patterns for particular products and services [122] will be different than they were pre-COVID-19. Furthermore, we observed that the predominant methodologies of reviewed studies have been qualitative. Given the early stages of knowledge regarding the pandemic, qualitative assessments and the gathering of previous findings may be critical in moving these issues forward; however, as more information becomes available, we should expect more research based on empirical and quantitative methodologies. Again, due to the novelty of the topic and the paucity of data on how different countries are dealing with the COVID-19 pandemic, a significant percentage of the reviewed publications have not concentrated on the geographical contexts or individual country performance. Because many nations have seen varying levels of infection and have chosen different lockdown approaches to cope with the pandemic issue, we believe that a regionally focused strategy will become increasingly relevant. As a result, the post-COVID-19 situation will be quite different, depending on the region. Despite this scarcity of data, our analysis revealed that nations such as China, India, the United States, and the Middle East, which experienced the greatest problems during the COVID-19 pandemic, have garnered more study attention than any other country. As a result, we believe that our second hypothesis may be validated based on current facts; the concentration of research on the effects of the BCT on SCM differs according to the conditions of various countries in terms of the COVID-19 pandemic's consequences. The findings show that articles focusing on the sustainability context are few. When we explore the social and environmental issues in the context of the COVID-19 pandemic, this shortcoming is more apparent. The applicability of BCT to these challenges, particularly in the present and post-pandemic situations, is unclear, and the research literature has given them little consideration. However, based on the results, we may infer that our third hypothesis is also correct. Despite the fact that the previous studies cannot show the effects of BCT on all aspects of the sustainable factors regarding COVID-19 conditions, it is clear that these effects are significant in the areas of economics and waste management.

Last but not least, we have identified the gaps in the literature and provide research opportunities and suggestions for future studies.

7.2. Suggestions for Future Works

Based on our findings and outlook in this study, we highlight the following as immediate extensions for future studies.

According to our findings, there are substantial shortcomings in both the empirical and quantitative methods. Furthermore, in the matter of geographical contexts, not enough attention has been devoted to the utilization of different nations' achievements in using BCT to address COVID-19-related difficulties regarding SCM. Despite the fact that BCT are utilized in many sectors in Europe and the United States, there is still a scarcity of research related to these nations. Furthermore, the proportion of sustainability studies in the research literature is quite low, and these flaws are especially evident when it comes to environmental and social concerns.

We propose that researchers explore these issues further with a positive outlook because, by the expansion of these research areas, particularly those that concentrate on the recovery and resiliency of sustainable SCs [123–125], businesses and governments can be better prepared for critical situations and increase their chances of survival in post-crisis conditions. Overall, we hope that our research will assist academics and other stakeholders in gaining a better understanding of the current literature on SCM and the ramifications of the pandemic, identifying areas that need more research, and directing future studies.

Turbulent times call for bold and innovative solutions for all the stages of SCM, from product design and production to distribution, thence to consumption. Although the focus of this study was on BCT and SCM literature during the COVID-19 pandemic, supply-chain strategies and solutions will undoubtedly be impacted by the advent of enhanced technologies such as BCT, the Internet of Things (e.g., Chadha et al. [126]), and sustainability imperatives. To that end, we next plan to work on how intelligent technologies can be aligned with both SC strategies (the logistics of SCM) regarding the United Nations Sustainable Development Goals. Various further extensions on this broad topic could examine, among others, product returns management and sustainability (e.g., Ülkü and Gürler [127]) and the integrative logistics needed for sustainable consumption and production in developing and emerging countries e.g., [128].

Author Contributions: Conceptualization, M.N. and M.A.Ü.; methodology, M.N. and M.A.Ü.; software, M.N. and M.A.Ü.; validation, M.N. and M.A.Ü.; formal analysis, M.N. and M.A.Ü.; investigation, M.N. and M.A.Ü.; resources, M.N. and M.A.Ü.; data curation, M.N. and M.A.Ü.; writing—original draft preparation, M.N.; writing—review and editing, M.N. and M.A.Ü.; visualization, M.N.; supervision, M.A.Ü.; project administration, M.N. and M.A.Ü.; funding acquisition, M.A.Ü. All authors have read and agreed to the published version of the manuscript.

Funding: This research was partially funded by the CRSSCA—Centre for Research in Sustainable Supply Chain Analytics, Dalhousie University, Canada. Grant number: CRSSCA#49100-20003-SCR.

Data Availability Statement: No real data is used in this study.

Acknowledgments: We thank the Editors and anonymous reviewers for their constructive comments.

Conflicts of Interest: The authors declare no conflict of interest.

Appendix A. Descriptive Analysis of Reviewed Studies

Table A1. Description reviewed articles on COVID-19 pandemic in supply chain and blockchain technologies deployment.

References	Research Description
Acioli et al. [77]	Analyzed the influences of BCT on SC operation in the COVID-19 period. The findings indicate that Industry 4.0 may create issues such as societal inequities connected to man's position in the global economy by replacing real employees with machinery.
Agarwal et al. [78]	Discussed the adoption of SC enablers to achieve greater efficiency against the COVID-19 pandemic interruption. The findings explore that among the characteristics of a resilient SC, trackability, and traceability of products get the closest attention.
Ahmad et al. [112]	Investigated decentralized BCTs to automate forward SC operations for the COVID-19 medical equipment. They offer a security study utilizing Ethereum algorithms to validate the dependability of smart contracts, and they explore solutions using cost analysis.
Akhigbe et al. [79]	Discussed the Internet of Things (IoT) technologies for livestock SC in the COVID-19 situation using a systematic review methodology. The findings show that there are enough advanced technology infrastructures to drive IoT for a variety of management objectives in BCT.
Bahn et al. [57]	Investigated the digitalization for sustainable agricultural-food systems in the COVID-19 conditions. The findings reveal that digital agriculture adoption is still in its early stages, with high-value agricultural output focusing on national consumers in Gulf nations and export markets in developing nations leading the way.
Bakalis et al. [100]	The findings show that technologies like IoT and BC are becoming essential in the risk management the food products, and the infrastructure of these technologies are closely related to key criteria like traceability, transparency, efficiency, and product quality to control COVID-19 circumstances.
Bamakan et al. [117]	The findings demonstrate that BCT might provide obvious advantages such as serialization, tracking, protecting IoT devices and smart contracts in pharmaceutical and medical projects.
Bekrar et al. [115]	Investigated the digitalizing benefits in the transportation sector facing the COVID-19. The article looked at BCT in various aspects, particularly as a permanent and trustworthy database, a monitoring tool, a smart contract function, and digital signatures.
Betcheva et al. [58]	Explored the SC thinking in the healthcare industry by considering the COVID-19 pandemic. The results show that the complexity in managing healthcare SCs offers opportunities for using technologies that cover visibility and transparency to develop various areas.
Bumblauskas et al. [101]	Investigated a BC use case in food distribution during the COVID-19 outbreak. The findings indicate that by building visible and transparent food SCs, stakeholders may obtain the required data for deciding on buying food products and supporting businesses.
Butt [59]	Explored the link between additive manufacturing and Industry 4.0 technologies considering the COVID-19 pandemic. The findings show that deploying technologies allow allows businesses to adapt to consumer demands more quickly, which will hasten the transition to smart manufacturing.
Chamola et al. [13]	Discussed the utilization of technologies like the BC, AI, and IoT to reduce the influences of the COVID-19 by a comprehensive review.

Table A1. *Cont.*

References	Research Description
Chowdhury et al. [60]	Reviewed the studies on the COVID-19 pandemic in SCs. The findings indicate a scarcity of research that is both empirically constructed and conceptually based. This study also presents research directions for future studies.
Choi [27]	Investigated the impacts of logistics technologies on transforming the static service operations to the dynamic home mobile service operations in the COVID-19 outbreak. According to the results, by applying BCT, operations could be more transparent, secure, and traceable.
Choi [102]	Explored the risk analysis in the logistics system during and after the COVID-19 pandemic. The findings indicate that the use of BC might be an efficient solution for several issues such as the need for remote working, enhancing transparency, and support the visibility of the logistics systems.
Choi [98]	Explored a framework for fighting against the COVID-19. The results show that using AI and BCT may enhance the traceability of patients suspected of infection, and by creating safe digital processes, you can make elections and voting easier.
Cordeiro et al. [61]	through a bibliometric analysis and systematic literature review, flaws in SCs may be used to develop public policies that increase resiliency, particularly during the COVID-19.
de Sousa Jabbour et al. [23]	Investigated sustainability of SCs during the COVID-19 pandemic. They presented a framework on four principles related to engineering, collaboration, agility, and culture. Findings show that technologies, especially BCT, play an influential role in the SC agility and risk management culture.
Di Vaio et al. [85]	Discussed artificial intelligence (AI) in the agricultural-food systems by the systematic literature review. The findings suggest that focusing on digital technologies, IoT, and BC to realize traceability of SCs, especially during the COVID-19 pandemic, might be a way to make this industry sector more robust.
Dolgui and Ivanov [116]	Discussed articles in the special issue utilizing various techniques, as well as compiling the latest findings. The BC, SC robustness, ripple effect, big data, and digital systems are all examples.
Dutta et al. [86]	Investigated BCT in SC operations by considering the COVID-19 impacts. They looked at a number of industries that may benefit from BCT because of improved visibility and business process management. In addition, a future study plan for this key developing research field is created.
Etemadi et al. [108]	The findings cover a variety of topics, including BC's potential for privacy and security problems, smart contract security, fraud monitoring, and tracking database systems to assure food safety and security.
Finkenstadt and Handfield [80]	The findings show that visibility and velocity are the most important characteristics for facilitating vital judgment reliability in the health care sector.
Fusco et al. [62]	Discussed the importance of BC in health care, in particular in terms of COVID19-safe clinical practice. Findings indicate that BC may be utilized in a new process with specific attention to risk management.
Golan et al. [20]	Discussed literature on SC robustness and linkages to sectors like transportation considering the COVID-19's circumstances. BC was presented as a mechanism for visibility between SC phases, and it may be required to prevent SC network failure.
Guo et al. [63]	The findings demonstrate that by utilizing BCT, SMEs may be able to successfully adapt to public crises (such as the COVID-19) as a result of digitalization.
Gurbuz and Ozkan [87]	Findings show that agriculture and food sectors have to adopt innovative approaches such as using modern BCT quickly to face the post-crisis uncertainty.
Iftekhar et al. [103]	Investigated BCT in the food sector for disruptive situations like the COVID-19. Findings show that the deployment of BCT in the food industry could help to provide more transparency and prevent potential food safety hazards.

Table A1. *Cont.*

References	Research Description
Ivanov [7]	Investigated SC resilience by lean thinking during the COVID-19 outbreak. The findings provide a paradigm that connects several aspects of successful robustness and enables the efficient use of resilience abilities in value generation.
Ivanov and Das [88]	The results show that the role of technologies in manufacturing the goods is significant concerning the capacity flexibility and product variety during the COVID-19 pandemic.
Ivanov and Dolgui [18]	The findings add to SC risks management studies by improving predictive and reactive choices to make use of the benefits of SC visibility and business sustainability in global enterprises.
Ivanov and Dolgui [5]	Explored the viability of intertwined supply networks (ISNs) for COVID-19. The results show that SC networks are moving towards the intertwined systems in which the visibility and digitalization technologies will be used as much as possible.
Kalla et al. [113]	Findings show that BCT is crucial in building a more robust SC and provides a high level of access restrictions and automation through intelligently designed contracts to build a reliable environment, particularly in the COVID-19 pandemic.
Karmaker et al. [64]	Findings indicate that monetary assistance from the government and SC partners is needed to face the imminent impact of COVID-19 on SC sustainability.
Kazancoglu et al. [89]	Discussed implementation of BCT in food supply chains based on the COVID-19 disruptions. The results show that there are significant relations between governmental incentives and management and SC visibility.
Kemp et al. [104]	While increased disclosure rules may offer visibility for traders, policymakers, and some other parties, they suggest that the internalization of transparency standards is highly complicated.
Kovács and Falagara Sigala [81]	Explored the humanitarian SCs' opportunities to mitigate and overcome SC disruptions during the COVID-19 pandemic. The findings show that technological innovations, such as using BC to trace deliveries, might be a good solution for more resiliency.
Kumar [65]	The findings demonstrate that using BCT in a distributed SC like COVID-19 may considerably minimize wastes and fake demand.
Kumar et al. [99]	Discussed the applications of industry 4.0 in the healthcare sector in the case of the COVID-19 pandemic. The findings show that technologies such as BC can act as significant drivers to build trust and transparency, reducing the impact of identified challenges.
Kumar and Kumar Singh [66]	The findings indicate that deploying BCT could assist all the food and agricultural stakeholders in overcoming the uncertain business environment like COVID-19′s conditions.
Kumar and Pundir [90]	Analyzed the deployment of BCT and IoT enablers in the pharmaceutical SC confronting the COVID-19 pandemic. The proposed framework could enhance the visibility, transparency, and privacy of the medical SC.
Labaran and Hamma-Adama [67]	The findings indicate some barriers to BC adoption within the Nigerian pharmaceutical SC, including the degree of knowledge of BCT between participants and governmental authorities, are extremely weak.
Lin et al. [114]	The findings demonstrate that the BCT, such as smart contracts and traceability, has significant influences on developing agricultural applications, and it generates a more productive food SC, particularly for post-COVID-19′s circumstances.
Lin et al. [82]	The findings suggest that attitudes and observed behavior controlling characteristics have a substantial and favorable impact on the desire to use BC food traceability technology to address COVID-19 disruptions.
Liu et al. [68]	The findings include a description of the study objectives, theoretical framework, and findings for emerging innovations such as BCT in operations and supply chain sustainability during the COVID-19 pandemic.

Table A1. *Cont.*

References	Research Description
Lohmer et al. [21]	Discussed the effects of the deployment of BCT on SC Resilience by considering the COVID-19 pandemic. The findings show that the BCT could intensify collaboration through smart contracts, and that sharing data using a BC solution could be useful on the disruption duration.
Memon et al. [83]	Investigated the COVID-19 situation in China and India, looking deeper into the pandemic's influence on the food and beverage industry, as well as exploring the programs and initiatives implemented for more resiliency.
Montecchi et al. [105]	Investigated the existing research on supply chain transparency. The findings provide a framework for academics to use in further research and for operators to use in their methods for new challenges that face supply chains concerning the COVID-19′ conditions.
Nandi et al. [91]	The findings from use cases demonstrate that BCT can help improve the global economy during the COVID-19 pandemic by supporting SC monitoring, traceability, and reactivity.
Nandi et al. [92]	The authors looked into how businesses might improve their flexibility and digitalization by utilizing BCT resources, During COVID-19. The findings propose a paradigm for further evaluating the link between SC resiliency and company capabilities.
Narayanamurthy and Tortorella [69]	Discussed the influences of the COVID-19 outbreak on employee performance. Findings show that industry 4.0 technologies, particularly the BCT, moderate the enhancement of employee performance.
Papadopoulos et al. [70]	Investigated the digitalization of companies during extreme and global disruptions. The findings show that digital technologies could be useful for small and intermediate enterprises to keep maintaining their activities during and after the COVID-19 pandemic.
Paul and Chowdhury [3]	Devised a production recovery strategy for essential goods like healthcare products confronting the COVID-19 outbreak. Findings show that BCT may help with the recovery process and that managers can use these technologies throughout the revival period.
Pillai and Mohan [106]	Discussed the blockchain usage in SCs operations in the contexts of the COVID-19 pandemic. The research concentrates on the function of BCT as a solution in a public distribution system for enhancing transparency and visibility across SC stakeholders.
Platt et al. [93]	Through a comprehensive review analysis, the authors looked at the use of BCT for improving digital contact tracking and reporting. The findings suggest that BCT may have a greater impact on public health in fields other than contact tracing.
Quayson et al. [71]	The study investigates how digitalization might protect the most fragile members of SC from catastrophic shocks, particularly for post-COVID-19′ circumstances.
Remko [22]	The results implicate that BCT deployment for enhancing visibility and transparency is an effective solution for improving SC resilience during and post COVID-19.
Rowan and Galanakis [109]	The findings indicate that the usage of BCT has the potential to significantly improve safety and security, and could also offer authenticity and traceability for agricultural and food SC for confronting after COVID-19 pandemic.
Sharma et al. [14]	The findings suggest that technologies like AI, automation, BCT, and deep learning might be critical for improving visibility and efficiency throughout the SC.
Sharma et al. [72]	Explored the priorities for retail SCs to integrate the operational activities for the post-COVID-19 period. According to findings, to mitigate the risks posed by COVID-19, organizations could leverage new technologies like BC.
Siriwardhana et al. [73]	Addressed new approaches like BCT and IoT in the retail sector, working remotely, and smart manufacturing considering the COVID-19's conditions.
Sodhi and Tang [107]	Explored SCM challenges for extreme conditions like the COVID-19 pandemic. The paper list research opportunities for SCM in extreme conditions, such as the usage of new technologies like BC to upgrade SC capacity in distribution conditions.

Table A1. *Cont.*

References	Research Description
Starr et al. [74]	Developed a digital foundation for the real estate industry by using Industry 4.0 technologies. The paper provides a primer on how BCT can embrace the rapid changes after the COVID-19 situation.
Sufian et al. [75]	Reviewed different methodologies in smart manufacturing by evaluating the newest trends of BCT. The findings show that the proposed plan may be used as a practical tool to fill the gap between advanced technologies and their industrial applications.
Tasnim [94]	Discussed a theoretical review of global food SC disruption considering the COVID-19 pandemic and digitalization by BCT. The results show that visibility and traceability have a foremost role in firms' SCM.
Xu et al. [95]	Proposed a theoretical framework that comprises four scenarios in the context of BCT for the implementation in coastal constructions considering the COVID-19′ circumstances.
Yadav et al. [110]	Explored the usage of BCT and IoT for developing the sustainability of agricultural-food SC under epidemic outbreaks such as COVID-19. This paper guides the organization's managers in their strategic planning based on digitalization enablers.
Yang et al. [84]	The authors conceptualize SC visibility as a mechanistic control to enhance supply chain resilience to manage crises like the COVID-19 pandemic.
Yeganeh [76]	The study found that the key changes triggered by the COVID-19 pandemic, notably the growing impact of large technologies, are a culmination of the upheavals of the previous decades, resulting in a new sort of globalization defined by extensive accessibility and intangible value.
Yousif et al. [96]	Discussed IoT and BCT during and beyond COVID-19 through a comprehensive review. The findings outline potential research directions for next-generation IoT and BCT applications that could improve the SCs' performance.
Zhang et al. [111]	Examined the history and prospects of operations management research. According to the findings, companies need to develop three essential capacities: connection, transparency, and consistency, to attain operating agility, resiliency, and viability in the age of Industry 4.0 age.
Zouari et al. [97]	Explored digitalization impacts on the SC resilience confronting the COVID-19 conditions. According to the authors, the level of digitalization and the deployment of digital technologies have a favorable influence on SC robustness.

Table A2. Summary of the reviewed articles based on the blockchain technology, methodologies, industry, geographical context, and sustainability issues. The sign √ relates the content of the paper with the factors considered.

References	Blockchain Technologies				Methodologies										Industry Sector									Geographical Classification											Sustainability Context				
	Digitalization	Visibility	Transparency	Smart Contracts	Explanatory	Conceptual	Case Study	Simulation	Mathematical	Survey	Interview	Literature Review	Decision Making	Statistical	Food and Agriculture	Health Care	Retail	Real State	Construction	Automotive	Transportation	Mining	Multiple Industries	Not Specified	India	China	US	Europe	Australia	Africa	Hong Kong	Middle East	Bangladesh	Multiple Countries	Not Specified	Waste Issues	Environmental	Economic Issues	Social Issues
Acioli et al. [77]		√										√												√											√				√
Agarwal et al. [78]		√											√							√					√														
Ahmad et al. [112]				√					√							√																			√	√			
Akhigbe et al. [79]		√										√			√																				√				
Bahn et al. [57]	√				√										√															√		√						√	
Bakalis et al. [100]		√	√									√			√																				√				
Bamakan et al. [117]	√	√	√	√			√					√				√																√				√			
Bekrar et al. [115]	√	√		√								√									√														√				
Betcheva et al. [58]	√						√									√																		√					
Bumblauskas et al. [101]		√	√				√								√												√												
Butt [59]	√				√																			√											√				
Chamola et al. [13]	√											√											√											√					
Chowdhury et al. [60]	√											√											√											√					
Choi [27]		√	√						√												√										√								
Choi [102]		√	√		√																			√											√				
Choi [98]	√	√			√																			√											√				
Cordeiro et al. [61]	√											√											√												√				
de Sousa Jabbour et al. [23]	√	√				√																		√											√			√	
Di Vaio et al. [85]	√	√										√			√																				√			√	
Dolgui and Ivanov [116]	√		√	√	√																			√											√				
Dutta et al. [86]	√	√										√											√												√				
Etemadi et al. [108]	√	√	√									√												√											√				
Finkenstadt and Handfield [80]		√				√										√											√												

Table A2. *Cont.*

References	Blockchain Technologies				Methodologies										Industry Sector										Geographical Classification											Sustainability Context				
	Digitalization	Visibility	Transparency	Smart Contracts	Explanatory	Conceptual	Case Study	Simulation	Mathematical	Survey	Interview	Literature Review	Decision Making	Statistical	Food and Agriculture	Health Care	Retail	Real State	Construction	Automotive	Transportation	Mining	Multiple Industries	Not Specified	India	China	US	Europe	Australia	Africa	Hong Kong	Middle East	Bangladesh	Multiple Countries	Not Specified	Waste Issues	Environmental	Economic Issues	Social Issues
Fusco et al. [62]	√												√			√																			√				
Golan et al. [20]		√										√											√												√				
Guo et al. [63]	√									√													√			√													
Gurbuz and Ozkan [87]	√	√								√				√	√																				√				
Iftekhar et al. [103]		√	√			√									√																				√				
Ivanov [7]		√				√																		√											√				
Ivanov and Das [88]	√	√						√																√										√					
Ivanov and Dolgui [18]	√	√				√																		√															
Ivanov and Dolgui [5]	√	√	√						√						√																				√				
Kalla et al. [113]	√			√	√																			√											√				
Karmaker et al. [64]	√								√															√									√					√	
Kazancoglu et al. [89]	√	√											√		√																				√	√		√	
Kemp et al. [104]		√	√				√				√											√												√		√			
Kovács and Falagara Sigala [81]		√			√											√					√														√		√		
Kumar [65]	√								√						√										√														
Kumar et al. [99]			√									√				√	√																		√				
Kumar and Kumar Singh [66]	√												√		√										√														
Kumar and Pundir [90]	√	√										√				√																			√				
Labaran and Hamma-Adama [67]	√										√					√														√									
Lin et al. [114]		√		√						√					√											√													
Lin et al. [82]		√				√				√					√											√													
Liu et al. [68]	√				√																			√											√				
Lohmer et al. [21]				√				√																√											√				
Memon et al. [83]		√			√									√	√										√	√													

Table A2. *Cont.*

References	Blockchain Technologies				Methodologies										Industry Sector									Geographical Classification											Sustainability Context				
	Digitalization	Visibility	Transparency	Smart Contracts	Explanatory	Conceptual	Case Study	Simulation	Mathematical	Survey	Interview	Literature Review	Decision Making	Statistical	Food and Agriculture	Health Care	Retail	Real State	Construction	Automotive	Transportation	Mining	Multiple Industries	Not Specified	India	China	US	Europe	Australia	Africa	Hong Kong	Middle East	Bangladesh	Multiple Countries	Not Specified	Waste Issues	Environmental	Economic Issues	Social Issues
Montecchi et al. [105]		√	√									√												√											√				
Nandi et al. [91]	√	√			√																		√												√				
Nandi et al. [92]	√	√			√																			√											√			√	
Narayanamurthy and Tortorella [69]	√									√				√										√	√														√
Papadopoulos et al. [70]	√				√																			√											√				
Paul and Chowdhury [3]	√								√							√													√										
Pillai and Mohan [106]		√	√		√										√										√														
Platt et al. [93]	√	√			√											√																			√				
Quayson et al. [71]	√				√										√															√									
Remko [22]	√	√	√								√													√			√												
Rowan and Galanakis [109]	√	√	√		√										√													√								√	√	√	
Sharma et al. [14]	√	√	√	√	√																			√											√				
Sharma et al. [72]	√								√				√				√																		√				
Siriwardhana et al. [73]	√				√																			√											√				
Sodhi and Tang [107]		√	√		√																			√											√				
Starr et al. [74]	√					√											√	√																	√				
Sufian et al. [75]	√					√						√												√											√				
Tasnim [94]	√	√			√										√																				√				
Xu et al. [95]	√	√				√					√								√												√								
Yadav et al. [110]	√	√	√										√		√																				√	√	√	√	
Yang et al. [84]		√								√													√			√					√								
Yeganeh [76]	√										√													√											√			√	√
Yousif et al. [96]	√	√										√				√					√														√				
Zhang et al. [111]	√	√	√		√																			√											√				
Zouari et al. [97]	√	√											√											√				√											

References

1. Hudecheck, M.; Sirén, C.; Grichnik, D.; Wincent, J. How companies can respond to the Coronavirus. *MIT Sloan Manag. Rev.* **2020**. Available online: https://sloanreview.mit.edu/article/how-companies-can-respond-to-the-coronavirus/ (accessed on 29 September 2020).
2. Gunessee, S.; Subramanian, N. Ambiguity and its coping mechanisms in supply chains lessons from the COVID-19 pandemic and natural disasters. *Int. J. Oper. Prod. Manag.* **2020**, *40*, 1201–1223. [CrossRef]
3. Paul, S.K.; Chowdhury, P. A production recovery plan in manufacturing supply chains for a high-demand item during COVID-19. *Int. J. Phys. Distrib. Logist. Manag.* **2020**, *51*, 104–125. [CrossRef]
4. Sodhi, M.S. Natural disasters, the economy and population vulnerability as a vicious cycle with exogenous hazards. *J. Oper. Manag.* **2016**, *45*, 101–113. [CrossRef]
5. Ivanov, D.; Dolgui, A. Viability of intertwined supply networks: Extending the supply chain resilience angles towards survivability. A position paper motivated by COVID-19 outbreak. *Int. J. Prod. Res.* **2020**, *58*, 2904–2915. [CrossRef]
6. Panday, A. Virus Outbreak Drives down Automobile Sales in March, Livemint. 2020. Available online: https://www.livemint.com/companies/news/automakers-hit-hard-bycovid-19%-march-sales-decline-over-50-11585749547879.html (accessed on 20 May 2021).
7. Ivanov, D. Lean resilience: AURA (Active Usage of Resilience Assets) framework for post-COVID-19 supply chain management. *Int. J. Logist. Manag.* **2021**. ahead-of-print. [CrossRef]
8. ACEA. *Interactive Map: Production Impact of COVID-19 on the European Auto Industry, 2020.* Available online: https://www.acea.be/news/article/interactive-map-production-impact-of-COVID-19-on-the-european-auto-industry (accessed on 1 July 2021).
9. IEA. *"Global Energy Review 2020," Report, International Energy Agency, Paris, 2020.* Available online: https://www.iea.org/countries/ukraine (accessed on 10 September 2020).
10. Guan, D.; Wang, D.; Hallegatte, S.; Davis, S.J.; Huo, J.; Li, S.; Bai, Y.; Lei, T.; Xue, Q.; Coffman, D.M.; et al. Global supply-chain effects of COVID-19 control measures. *Nat. Hum. Behav.* **2020**, *4*, 577–587. [CrossRef]
11. Szymkowski, S. COVID-19 Shut Down 93% of All US Auto Production. Roadshow. 2020. Available online: https://www.cnet.com/roadshow/news/COVID-19-shut-down-us-auto-production-coronavirus/ (accessed on 1 April 2021).
12. Amankwah-Amoah, J. Stepping up and stepping out of COVID-19: New challenges for environmental sustainability policies in the global airline industry. *J. Clean. Prod.* **2020**, *271*, 123000. [CrossRef]
13. Chamola, V.; Hassija, V.; Gupta, V.; Guizani, M. A comprehensive review of the COVID-19 pandemic and the role of IoT, drones, AI, blockchain, and 5G in managing its mpact. *IEEE Access* **2020**, *8*, 90225–90265. [CrossRef]
14. Sharma, A.; Adhikary, A.; Borah, S.B. COVID-19's Impact on Supply Chain Decisions: Strategic Insights for NASDAQ 100 Firms using Twitter Data. *J. Bus. Res.* **2020**, *117*, 443–449. [CrossRef]
15. Goel, R.K.; Saunoris, J.W.; Goel, S.S. Supply Chain Reliability and International Economic Growth: Impacts of Disruptions like COVID-19. CESifo Working Paper 2020, No. 8294. Available online: https://ssrn.com/abstract=3603829 (accessed on 1 April 2021).
16. Hinge, G.; Muksor, A.; Goyal, M.K.; Goyal, R.K.; Goyal, M. Integrated risk of virus outbreaks. In *Integrated Risk of Pandemic: COVID-19 Impacts, Resilience and Recommendations. Disaster Resilience and Green Growth*; Indian Institute of Technology, National Institute of Disaster Management, Eds.; Springer: Singapore, 2020. [CrossRef]
17. Ranney, M.L.; Griffeth, V.; Jha, A.K. Critical supply shortages—The need for ventilators and personal protective equipment during the COVID-19 pandemic. *N. Engl. J. Med.* **2020**, *382*, e41. [CrossRef] [PubMed]
18. Ivanov, D.; Dolgui, A. A digital supply chain twin for managing the disruption risks and resilience in the era of Industry 4.0. *Prod. Plan. Control.* **2020**, *32*, 1–14. [CrossRef]
19. Fasan, M.; Soerger Zaro, E.; Soerger Zaro, C.; Porco, B.; Tiscini, R. An empirical analysis: Did green supply chain management alleviate the effects of COVID-19? *Bus. Strategy Environ.* **2021**, *30*, 2702–2712. [CrossRef] [PubMed]
20. Golan, M.S.; Jernegan, L.H.; Linkov, I. Trends and applications of resilience analytics in supply chain modeling: Systematic literature review in the context of the COVID-19 pandemic. *Environ. Syst. Decis.* **2020**, *40*, 222–243. [CrossRef] [PubMed]
21. Lohmer, J.; Bugert, N.; Lasch, R. Analysis of resilience strategies and ripple effect in blockchain-coordinated supply chains: An agent-based simulation study. *Int. J. Prod. Econ.* **2020**, *228*, 107882. [CrossRef] [PubMed]
22. Remko, V.H. Research opportunities for a more resilient post-COVID-19 supply chain–closing the gap between research findings and industry practice. *Int. J. Oper. Prod. Manag.* **2020**, *40*, 341–355. [CrossRef]
23. De Sousa Jabbour, A.B.L.; Jabbour, C.J.C.; Hingley, M.; Vilalta-Perdomo, E.L.; Ramsden, G.; Twigg, D. Sustainability of supply chains in the wake of the coronavirus (COVID-19/SARS-CoV-2) pandemic: Lessons and trends. *Mod. Supply Chain. Res. Appl.* **2020**, *2*, 117–122. [CrossRef]
24. Sarkis, J. Supply chain sustainability: Learning from the COVID-19 pandemic. *Int. J. Oper. Prod. Manag.* **2020**, *41*, 63–73. [CrossRef]
25. Tranfield, D.; Denyer, D.; Smart, P. Towards a methodology for developing evidence-informed management knowledge by means of systematic review. *Br. J. Manag.* **2003**, *14*, 207–222. [CrossRef]
26. Wamba, S.F.; Queiroz, M.M. Blockchain in the operations and supply chain management: Benefits, challenges, and future research opportunities. *Int. J. Inf. Manag.* **2020**, *52*, 102064. [CrossRef]

27. Choi, T.M. Innovative "bring-service-near-your-home" operations under Corona-virus (COVID-19/SARS-CoV-2) outbreak: Can logistics become the messiah? *Transp. Res. Part E Logist. Transp. Rev.* **2020**, *140*, 101961. [CrossRef]
28. Bager, S.L.; Lambin, E.F. Sustainability strategies by companies in the global coffee sector. *Bus. Strategy Environ.* **2020**, *2*, 3555–3570. [CrossRef]
29. Chang, S.E.; Chen, Y.C.; Lu, M.F. Supply chain re-engineering using blockchain technology: A case of smart contract-based tracking process. *Technol. Forecast. Soc. Chang.* **2019**, *144*, 1–11. [CrossRef]
30. Rodríguez-Espíndola, O.; Chowdhury, S.; Beltagui, A.; Albores, P. The potential of emergent disruptive technologies for humanitarian supply chains: The integration of blockchain, Artificial Intelligence and 3D printing. *Int. J. Prod. Res.* **2020**, *58*, 4610–4630. [CrossRef]
31. Wright, A.; De Filippi, P. Decentralized Blockchain Technology and the Rise of Lex Cryptographia. 2015. Available online: https://ssrn.com/abstract=2580664 (accessed on 1 May 2020).
32. Baumers, M.; Holweg, M. On the economics of additive manufacturing: Experimental indings. *J. Oper. Manag.* **2019**, *65*, 794–809. [CrossRef]
33. Babich, V.; Hilary, G. OM Forum—Distributed ledgers and operations: What operations management researchers should know about blockchain technology. *Manuf. Serv. Oper. Manag.* **2020**, *22*, 223–240. [CrossRef]
34. Cai, Y.J.; Choi, T.M.; Zhang, J. Platform supported supply chain operations in the blockchain era: Supply contracting and moral hazards. *Decis. Sci.* **2020**, *52*, 866–892. [CrossRef]
35. Bai, C.A.; Cordeiro, J.; Sarkis, J. Blockchain technology: Business, strategy, the environment, and sustainability. *Bus. Strategy Environ.* **2020**, *29*, 321–322. [CrossRef]
36. Wang, Y.; Han, J.H.; Beynon-Davies, P. Understanding blockchain technology for future supply chains: A systematic literature review and research agenda. *Supply Chain. Manag. Int. J.* **2019**, *24*, 62–84. [CrossRef]
37. Carter, C.R.; Rogers, D.S.; Choi, T.Y. Toward the theory of the supply chain. *J. Supply Chain. Manag.* **2015**, *51*, 89–97. [CrossRef]
38. Frohlich, M.T.; Westbrook, R. Arcs of integration: An international study of supply chain strategies. *J. Oper. Manag.* **2001**, *19*, 185–200. [CrossRef]
39. Nakasumi, M. Information sharing for supply chain management based on blockchain technology. In Proceedings of the 2017 IEEE 19th Conference on Business Informatics (CBI), Thessaloniki, Greece, 24–27 July 2017; Volume 1, pp. 140–149. [CrossRef]
40. Michelman, P. Seeing beyond the blockchain hype. *MIT Sloan Manag. Rev.* **2017**, *58*, 17–19.
41. Wang, J.; Wu, P.; Wang, X.; Shou, W. The outlook of blockchain technology for construction engineering management. *Front. Eng. Manag.* **2017**, *4*, 67–75. [CrossRef]
42. Casey, M.J.; Wong, P. Global supply chains are about to get better, thanks to blockchain. *Harv. Bus. Rev.* **2017**, *13*, 1–6.
43. Mansfield-Devine, S. Beyond Bitcoin: Using blockchain technology to provide assurance in the commercial world. *Comput. Fraud. Secur.* **2017**, *2017*, 14–18. [CrossRef]
44. Roeck, D.; Sternberg, H.; Hofmann, E. Distributed ledger technology in supply chains: A transaction cost perspective. *Int. J. Prod. Res.* **2020**, *58*, 2124–2141. [CrossRef]
45. Van Hoek, R. Exploring blockchain implementation in the supply chain. *Int. J. Oper. Prod. Manag.* **2019**, *39*, 829–859. [CrossRef]
46. Gurtu, A.; Johny, J. Potential of blockchain technology in supply chain management: A literature review. *Int. J. Phys. Distrib. Logist.* **2019**, *49*, 881–900. [CrossRef]
47. Tian, F. An agri-food supply chain traceability system for China based on RFID & blockchain technology. In Proceedings of the 13th International Conference on Service Systems and Service Management (ICSSSM), Kunming, China, 24–26 June 2016; pp. 1–6. [CrossRef]
48. Collomb, A.; Sok, K. Blockchain/distributed ledger technology (DLT): What impact on the financial sector? *Digiworld Econ. J.* **2016**, *3*, 93–111.
49. Saberi, S.; Kouhizadeh, M.; Sarkis, J.; Shen, L. Blockchain technology and its relationships to sustainable supply chain management. *Int. J. Prod. Res.* **2019**, *57*, 2117–2135. [CrossRef]
50. Francisco, K.; Swanson, D. The supply chain has no clothes: Technology adoption of blockchain for supply chain transparency. *Logistics* **2018**, *2*, 2. [CrossRef]
51. Van Engelenburg, S.; Janssen, M.; Klievink, B. Design of a software architecture supporting business-to-government information sharing to improve public safety and security. *J. Intell. Inf. Syst.* **2019**, *52*, 595–618. [CrossRef]
52. Bocek, T.; Rodrigues, B.B.; Strasser, T.; Stiller, B. Blockchains everywhere-a use-case of blockchains in the pharma supply chain. In Proceedings of the 2017 IFIP/IEEE Symposium on Integrated Network and Service Management (IM), Lisbon, Portugal, 8–12 May 2017; pp. 772–777. [CrossRef]
53. Dolgui, A.; Ivanov, D.; Potryasaev, S.; Sokolov, B.; Ivanova, M.; Werner, F. Blockchain-oriented dynamic modeling of smart contract design and execution in the supply chain. *Int. J. Prod. Res.* **2020**, *58*, 2184–2199. [CrossRef]
54. Omar, I.A.; Jayaraman, R.; Debe, M.S.; Salah, K.; Yaqoob, I.; Omar, M. Automating procurement contracts in the healthcare supply chain using blockchain smart contracts. *IEEE Access* **2021**, *9*, 37397–37409. [CrossRef]
55. Bryman, A. *Social Research Methods*, 4th ed.; Oxford University Press: New York, NY, USA, 2012.
56. Saunders, M.; Lewis, P.; Thornhill, A. *Research Methods for Business Students*; Pearson Education: London, UK, 2009.
57. Bahn, R.A.; Yehya, A.A.K.; Zurayk, R. Digitalization for Sustainable Agri-Food Systems: Potential, Status, and Risks for the MENA Region. *Sustainability* **2021**, *13*, 3223. [CrossRef]

58. Betcheva, L.; Erhun, F.; Jiang, H. Supply Chain Thinking in Healthcare: Lessons and Outlooks. *Manuf. Serv. Oper. Manag.* **2020**. ahead-of-print. [CrossRef]
59. Butt, J. Exploring the interrelationship between additive manufacturing and Industry 4.0. *Designs* **2020**, *4*, 13. [CrossRef]
60. Chowdhury, P.; Paul, S.K.; Kaisar, S.; Moktadir, M.A. COVID-19 pandemic related supply chain studies: A systematic review. *Transp. Res. Part E Logist. Transp. Rev.* **2021**, *148*, 102271. [CrossRef]
61. Cordeiro, M.C.; Santos, L.; Angelo, A.C.M.; Marujo, L.G. Research directions for supply chain management in facing pandemics: An assessment based on bibliometric analysis and systematic literature review. *Int. J. Logist. Res. Appl.* **2021**. ahead-of-print. [CrossRef]
62. Fusco, A.; Dicuonzo, G.; Dell'Atti, V.; Tatullo, M. Blockchain in Healthcare: Insights on COVID-19. *Int. J. Environ. Res. Public Health* **2020**, *17*, 7167. [CrossRef]
63. Guo, H.; Yang, Z.; Huang, R.; Guo, A. The digitalization and public crisis responses of small and medium enterprises: Implications from a COVID-19 survey. *Front. Bus. Res. China* **2020**, *14*, 1–25. [CrossRef]
64. Karmaker, C.L.; Ahmed, T.; Ahmed, S.; Ali, S.M.; Moktadir, M.A.; Kabir, G. Improving supply chain sustainability in the context of COVID-19 pandemic in an emerging economy: Exploring drivers using an integrated model. *Sustain. Prod. Consum.* **2021**, *26*, 411–427. [CrossRef] [PubMed]
65. Kumar, A. Improvement of public distribution system efficiency applying blockchain technology during pandemic outbreak (COVID-19). *J. Humanit. Logist. Supply Chain. Manag.* **2020**, *11*, 1–28. [CrossRef]
66. Kumar, P.; Kumar Singh, R. Strategic framework for developing resilience in Agri-Food Supply Chains during COVID 19 pandemic. *Int. J. Logist. Res. Appl.* **2021**. ahead-of-print. [CrossRef]
67. Labaran, M.J.; Hamma-Adama, M. The Nigerian pharmaceutical supply chain: Blockchain adoption, counterfeit drugs and successful deployment of COVID-19 vaccine in Nigeria. *J. Sci. Res. Rep.* **2021**, *27*, 20–36. [CrossRef]
68. Liu, Y.; Zhu, Q.; Seuring, S. New technologies in operations and supply chains: Implications for sustainability. *Int. J. Prod. Econ.* **2020**, *229*, 107889. [CrossRef]
69. Narayanamurthy, G.; Tortorella, G. Impact of COVID-19 outbreak on employee performance–moderating role of industry 4.0 base technologies. *Int. J. Prod. Econ.* **2021**, *234*, 108075. [CrossRef]
70. Papadopoulos, T.; Baltas, K.N.; Balta, M.E. The use of digital technologies by small and medium enterprises during COVID-19: Implications for theory and practice. *Int. J. Inf. Manag.* **2020**, *55*, 102192. [CrossRef] [PubMed]
71. Quayson, M.; Bai, C.; Osei, V. Digital inclusion for resilient post-COVID-19 supply chains: Smallholder farmer perspectives. *IEEE Eng. Manag. Rev.* **2020**, *48*, 104–110. [CrossRef]
72. Sharma, M.; Luthra, S.; Joshi, S.; Kumar, A. Accelerating retail supply chain performance against pandemic disruption: Adopting resilient strategies to mitigate the long-term effects. *J. Enterp. Inf. Manag.* **2021**. ahead-of-print. [CrossRef]
73. Siriwardhana, Y.; De Alwis, C.; Gür, G.; Ylianttila, M.; Liyanage, M. The fight against the COVID-19 pandemic with 5G technologies. *IEEE Eng. Manag. Rev.* **2020**, *48*, 72–84. [CrossRef]
74. Starr, C.W.; Saginor, J.; Worzala, E. The rise of PropTech: Emerging industrial technologies and their impact on real estate. *J. Prop. Invest. Financ.* **2020**, *39*, 157–169. [CrossRef]
75. Sufian, A.T.; Abdullah, B.M.; Ateeq, M.; Wah, R.; Clements, D. Six-Gear Roadmap towards the Smart Factory. *Appl. Sci.* **2021**, *11*, 3568. [CrossRef]
76. Yeganeh, H. Emerging social and business trends associated with the COVID-19 pandemic. *Crit. Perspect. Int. Bus.* **2021**, *17*, 188–209. [CrossRef]
77. Acioli, C.; Scavarda, A.; Reis, A. Applying Industry 4.0 technologies in the COVID–19 sustainable chains. *Int. J. Product. Perform. Manag.* **2021**, *70*, 988–1016. [CrossRef]
78. Agarwal, N.; Seth, N.; Agarwal, A. Modeling supply chain enablers for effective resilience. *Contin. Resil. Rev.* **2020**, *2*, 97–110. [CrossRef]
79. Akhigbe, B.I.; Munir, K.; Akinade, O.; Akanbi, L.; Oyedele, L.O. IoT technologies for livestock management: A review of present status, opportunities, and future trends. *Big Data Cogn. Comput.* **2021**, *5*, 10. [CrossRef]
80. Finkenstadt, D.J.; Handfield, R. Blurry vision: Supply chain visibility for personal protective equipment during COVID-19. *J. Purch. Supply Manag.* **2021**, *27*, 100689. [CrossRef]
81. Kovács, G.; Falagara Sigala, I. Lessons learned from humanitarian logistics to manage supply chain disruptions. *J. Supply Chain. Manag.* **2021**, *57*, 41–49. [CrossRef]
82. Lin, X.; Chang, S.C.; Chou, T.H.; Chen, S.C.; Ruangkanjanases, A. Consumers' Intention to Adopt Blockchain Food Traceability Technology towards Organic Food Products. *Int. J. Environ. Res. Public Health* **2021**, *18*, 912. [CrossRef] [PubMed]
83. Memon, S.U.R.; Pawase, V.R.; Pavase, T.R.; Soomro, M.A. Investigation of COVID-19 Impact on the Food and Beverages Industry: China and India Perspective. *Foods* **2021**, *10*, 1069. [CrossRef]
84. Yang, J.; Xie, H.; Yu, G.; Liu, M. Antecedents and consequences of supply chain risk management capabilities: An investigation in the post-coronavirus crisis. *Int. J. Prod. Res.* **2021**, *59*, 1573–1585. [CrossRef]
85. Di Vaio, A.; Boccia, F.; Landriani, L.; Palladino, R. Artificial intelligence in the agri-food system: Rethinking sustainable business models in the COVID-19 scenario. *Sustainability* **2020**, *12*, 4851. [CrossRef]
86. Dutta, P.; Choi, T.M.; Somani, S.; Butala, R. Blockchain technology in supply chain operations: Applications, challenges, and research opportunities. *Transp. Res. Part E Logist. Transp. Rev.* **2020**, *142*, 102067. [CrossRef]

87. Gurbuz, I.B.; Ozkan, G. Transform or perish: Preparing the business for a postpandemic future. *IEEE Eng. Manag. Rev.* **2020**, *48*, 139–145. [CrossRef]
88. Ivanov, D.; Das, A. Coronavirus (COVID-19/SARS-CoV-2) and supply chain resilience: A research note. *Int. J. Integr. Supply Manag.* **2020**, *13*, 90–102. [CrossRef]
89. Kazancoglu, Y.; Pala, M.O.; Sezer, M.D.; Luthra, S.; Kumar, A. Drivers of implementing Big Data Analytics in food supply chains for transition to a circular economy and sustainable operations management. *J. Enterp. Inf. Manag.* **2021**. ahead-of-print. [CrossRef]
90. Kumar, S.; Pundir, A.K. Blockchain–internet of things (IoT) enabled pharmaceutical supply chain for COVID-19. In Proceedings of the NA International Conference on Industrial Engineering and Operations Management, Detroit, MI, USA, 10–14 August 2020.
91. Nandi, S.; Sarkis, J.; Hervani, A.; Helms, M. Do blockchain and circular economy practices improve post COVID-19 supply chains? A resource-based and resource dependence perspective. *Ind. Manag. Data Syst.* **2020**, *121*, 333–363. [CrossRef]
92. Nandi, S.; Sarkis, J.; Hervani, A.A.; Helms, M.M. Redesigning supply chains using blockchain-enabled circular economy and COVID-19 experiences. *Sustain. Prod. Consum.* **2021**, *27*, 10–22. [CrossRef]
93. Platt, M.; Hasselgren, A.; Román-Belmonte, J.M.; De Oliveira, M.T.; De la Corte-Rodríguez, H.; Olabarriaga, S.D.; Rodríguez-Merchán, E.C.; Mackey, T.K. Test, Trace, and Put on the Blockchain?: A Viewpoint Evaluating the Use of Decentralized Systems for Algorithmic Contact Tracing to Combat a Global Pandemic. *JMIR Public Health Surveill.* **2021**, *7*, e26460. [CrossRef] [PubMed]
94. Tasnim, Z. Disruption in Global Food Supply Chain (FSCs) Due to Covid-19 Pandemic and Impact of Digitalization Through Blockchain Technology in FSCs Management. *Eur. J. Bus. Manag.* **2021**, *12*, 64–72. [CrossRef]
95. Xu, J.; Ye, M.; Lu, W.; Bao, Z.; Webster, C. A four-quadrant conceptual framework for analyzing extended producer responsibility in offshore prefabrication construction. *J. Clean. Prod.* **2021**, *282*, 124540. [CrossRef] [PubMed]
96. Yousif, M.; Hewage, C.; Nawaf, L. IoT Technologies during and beyond COVID-19: A Comprehensive Review. *Future Internet* **2021**, *13*, 105. [CrossRef]
97. Zouari, D.; Ruel, S.; Viale, L. Does digitalizing the supply chain contribute to its resilience? *Int. J. Phys. Distrib. Logist. Manag.* **2020**, *51*, 149–180. [CrossRef]
98. Choi, T.M. Fighting against COVID-19: What operations research can help and the sense-and-respond framework. *Ann. Oper. Res.* **2021**. ahead-of-print. [CrossRef]
99. Kumar, M.S.; Raut, R.D.; Narwane, V.S.; Narkhede, B.E. Applications of industry 4.0 to overcome the COVID-19 operational challenges. *Diabetes Metab. Syndr. Clin. Res. Rev.* **2020**, *14*, 1283–1289. [CrossRef] [PubMed]
100. Bakalis, S.; Valdramidis, V.; Argyropoulos, D.; Ahrne, L.; Chen, J.; Cullen, P.J.; Cummins, E.; Datta, A.K.; Emmanouilidis, C.; Foster, T.; et al. How COVID-19 changed our food systems and food security paradigms. *Curr. Res. Food Sci.* **2020**, *3*, 166–172. [CrossRef]
101. Bumblauskas, D.; Mann, A.; Dugan, B.; Rittmer, J. A blockchain use case in food distribution: Do you know where your food has been? *Int. J. Inf. Manag.* **2020**, *52*, 102008. [CrossRef]
102. Choi, T.M. Risk analysis in logistics systems: A research agenda during and after the COVID-19 pandemic. *Transp. Res. Part E Logist. Transp. Rev.* **2020**, *145*, 102190. [CrossRef]
103. Iftekhar, A.; Cui, X.; Hassan, M.; Afzal, W. Application of Blockchain and Internet of Things to ensure tamper-proof data availability for food safety. *J. Food Qual.* **2020**, *2020*, 5385207. [CrossRef]
104. Kemp, D.; Owen, J.R.; Lèbre, É. Tailings facility failures in the global mining industry: Will a 'transparency turn' drive change? *Bus. Strategy Environ.* **2021**, *30*, 122–134. [CrossRef]
105. Montecchi, M.; Plangger, K.; West, D. Supply chain transparency: A bibliometric review and research agenda. *Int. J. Prod. Econ.* **2021**, *238*, 108152. [CrossRef]
106. Pillai, S.V.; Vipin, H.; Mohan, A.; Shruthi, D.A. Infusing Blockchain in supply chain operations of a public distribution system. *J. Supply Chain. Manag. Syst.* **2020**, *9*, 1.
107. Sodhi, M.S.; Tang, C.S. Supply chain management for extreme conditions: Research opportunities. *J. Supply Chain. Manag.* **2021**, *57*, 7–16. [CrossRef]
108. Etemadi, N.; Borbon-Galvez, Y.; Strozzi, F.; Etemadi, T. Supply chain disruption risk management with Blockchain: A dynamic literature review. *Information* **2020**, *12*, 70. [CrossRef]
109. Rowan, N.J.; Galanakis, C.M. Unlocking challenges and opportunities presented by COVID-19 pandemic for cross-cutting disruption in agri-food and green deal innovations: Quo Vadis? *Sci. Total. Environ.* **2020**, *748*, 141362. [CrossRef] [PubMed]
110. Yadav, S.; Luthra, S.; Garg, D. Modeling Internet of Things (IoT)-driven global sustainability in multi-tier agri-food supply chain under natural epidemic outbreaks. *Environ. Sci. Pollut. Res.* **2021**, *28*, 16633–16654. [CrossRef] [PubMed]
111. Zhang, F.; Wu, X.; Tang, C.S.; Feng, T.; Dai, Y. Evolution of operations management research: From managing flows to building capabilities. *Prod. Oper. Manag.* **2020**, *29*, 2219–2229. [CrossRef]
112. Ahmad, R.W.; Salah, K.; Jayaraman, R.; Yaqoob, I.; Omar, M.; Ellahham, S. Blockchain-based forward supply chain and waste management for COVID-19 medical equipment and supplies. *IEEE Access* **2021**, *9*, 44905–44927. [CrossRef]
113. Kalla, A.; Hewa, T.; Mishra, R.A.; Ylianttila, M.; Liyanage, M. The role of blockchain to fight against COVID-19. *IEEE Eng. Manag. Rev.* **2020**, *48*, 85–96. [CrossRef]
114. Lin, W.; Huang, X.; Fang, H.; Wang, V.; Hua, Y.; Wang, J.; Yau, L. Blockchain technology in current agricultural systems: From techniques to applications. *IEEE Access* **2020**, *8*, 143920–143937. [CrossRef]

115. Bekrar, A.; El Cadi, A.A.; Todosijevic, R.; Sarkis, J. Digitalizing the closing-of-the-loop for supply chains: A transportation and Blockchain perspective. *Sustainability* **2021**, *13*, 2895. [CrossRef]
116. Dolgui, A.; Ivanov, D. Exploring supply chain structural dynamics: New disruptive technologies and disruption risks. *Int. J. Prod. Econ.* **2020**, *229*, 107886. [CrossRef]
117. Bamakan, S.M.H.; Moghaddam, S.G.; Manshadi, S.D. Blockchain-enabled pharmaceutical cold chain: Applications, key challenges, and future trends. *J. Clean. Prod.* **2021**, *302*, 127021. [CrossRef]
118. Centobelli, P.; Cerchione, R.; Ertz, M. Managing supply chain resilience to pursue business and environmental strategies. *Bus. Strategy Environ.* **2020**, *29*, 1215–1246. [CrossRef]
119. Flynn, B.; Cantor, D.; Pagell, M.; Dooley, K.; Azadegan, A. Managing supply chains beyond COVID-19: Preparing for the next global mega-disruption. *J. Supply Chain. Manag.* **2020**, *57*, 3–6. [CrossRef]
120. Grashuis, J.; Skevas, T.; Segovia, M.S. Grocery shopping preferences during the COVID-19 pandemic. *Sustainability* **2020**, *12*, 5369. [CrossRef]
121. Yu, Z.; Razzaq, A.; Rehman, A.; Shah, A.; Jameel, K.; Mor, R.S. Disruption in global supply chain and socio-economic shocks: A lesson from COVID-19 for sustainable production and consumption. *Oper. Manag. Res.* **2021**. ahead-of-print. [CrossRef]
122. Currie, B.A.; French, A.D.; Ülkü, M.A. Big data, sustainability, and consumer behavior: A supply chain framework. In *Big Data Analytics in Supply Chain Management: Theory and Applications*; Rahimi, I., Ed.; CRC Press, Taylor Francis Group: Boca Raton, FL, USA, 2021; pp. 109–132.
123. Wong, C.W.; Wong, C.Y.; Boon-itt, S. How does sustainable development of supply chains make firms lean, green and profitable? A resource orchestration perspective. *Bus. Strategy Environ.* **2018**, *27*, 375–388. [CrossRef]
124. Ülkü, M.A.; Engau, A. Sustainable supply chain analytics. In *Industry, Innovation and Infrastructure. Encyclopedia of the UN Sustainable Development Goals*; Leal Filho, W., Azul, A.M., Brandli, L., Lange Salvia, A., Wall, T., Eds.; Springer: Cham, Switzerland, 2021; pp. 1123–1134. [CrossRef]
125. Khan, S.A.R.; Razzaq, A.; Yu, Z.; Miller, S. Industry 4.0 and circular economy practices: A new era business strategies for environmental sustainability. *Bus. Strategy Environ.* **2021**. [CrossRef]
126. Chadha, S.S.; Ülkü, M.A.; Venkatadri, U. Freight delivery in a Physical Internet Supply Chain: An applied optimization model with peddling and shipment consolidation. *Int. J. Prod. Res.* **2021**. ahead-of-print. [CrossRef]
127. Ülkü, M.A.; Gürler, Ü. The impact of abusing return policies: A newsvendor model with opportunistic consumers. *Int. J. Prod. Econ.* **2018**, *203*, 124–133. [CrossRef]
128. Worku, A.; Ülkü, M.A. Analyzing the impact of the COVID-19 pandemic on vegetable market supply in Northwestern Ethiopia. *J. Agribus. Dev. Emerg. Econ. Int.* **2021**. ahead-of-print. [CrossRef]

MDPI
St. Alban-Anlage 66
4052 Basel
Switzerland
Tel. +41 61 683 77 34
Fax +41 61 302 89 18
www.mdpi.com

Sustainability Editorial Office
E-mail: sustainability@mdpi.com
www.mdpi.com/journal/sustainability

www.ingramcontent.com/pod-product-compliance
Lightning Source LLC
Chambersburg PA
CBHW062057220926
51850CB00010B/2796
* 9 7 8 3 0 3 6 5 3 7 3 1 3 *